Industry
Self-Regulation
and
Voluntary
Environmental
Compliance

Industry Self-Regulation and Voluntary Environmental Compliance

Al Iannuzzi, Jr.

CRC Press
Taylor & Francis Group
Boca Raton London New York

CRC Press is an imprint of the
Taylor & Francis Group, an **informa** business

CRC Press
Taylor & Francis Group
6000 Broken Sound Parkway NW, Suite 300
Boca Raton, FL 33487-2742

First issued in paperback 2020

© 2002 by Taylor & Francis Group, LLC
CRC Press is an imprint of Taylor & Francis Group, an Informa business

No claim to original U.S. Government works

ISBN 13: 978-0-367-57874-9 (pbk)
ISBN 13: 978-1-56670-570-7 (hbk)

Visit the Taylor & Francis Web site at
http://www.taylorandfrancis.com

and the CRC Press Web site at
http://www.crcpress.com

Cover illustration created by George Abe of Seattle, Washington.

Library of Congress Cataloging-in-Publication Data

Iannuzzi, Alphonse.
Industry self-regulation and voluntary environmental compliance / Alphonse Iannuzzi, Jr.
p. cm.
Includes bibliographical references and index.
ISBN 1-56670-5703-7
1. Environmental law—United States. 2. Environmental policy—United States. 3. Industrial policy—United States. 4. Environmental law—United States—Compliance costs.
I. Title.

KF3775 .I16 2001
344.73'046—dc21   2001046200

Library of Congress Card Number 2001046200

Foreword

This book is thoughtfully bullish about self-regulation to protect the environment. The author is a highly respected person from a highly respected company. While I am much more cautious than he, I am persuaded that self-regulation is worth a try, and here is why.

Environmental protection is a messy affair in the U.S., far more messy than in any other country — though arguably more successful as well. Regulators often implement rules without regard for business processes, which almost always leads to inefficiency and unnecessary expense. Compliance failures generate reams of lawsuits. The trust levels among industry, regulators, and environmental groups are low all the way around. And each player is very powerful.

But the funny thing is, regulation works. Flows of materials that are regulated go down, often in absolute terms, and always relative to production. Flows of unregulated substances go up. A recent study by my colleague Emily Matthews at the World Resources Institute revealed the effect of regulation on certain substances in the U.S. economy. She tracked the flows of dozens of materials over time. In case after case, the effect of regulation, or the absence of regulation, is evident in the prevalence of a substance in commerce. The case of CO_2 is interesting. CO_2 flows track economic growth, slowing with recessions, but essentially rising and rising. By weight, it accounts for nearly 95% of total flows through the economy. Yes, 95%! It is unregulated, and it goes up. If regulated, it would go down.

However, there are far too many substances and processes being introduced at too rapid a pace for regulation to keep up. Moreover, regulatory resources are scarce. As Dr. Iannuzzi points out, self-regulation would free up regulatory resources to focus on rogue actors or particularly vexing problems.

What is not so clear is whether self-regulation works. Dr. Iannuzzi argues compellingly that it does, offering four strong case studies. Self-regulation certainly could be a vital part of an overall environmental protection system, working to support regulation. When there is a steep price to pay for a certain protection, it would take a rare company or person to adopt it. It would take someone willing to flout the overwhelming pressures to improve financial performance. In these circumstances, it simply will not happen often enough to make a difference, and we need a regulator to protect the public against conflicting private interests. Hence, the challenge for self-regulation is not to supplant — but to complement — the existing regulatory system.

The remaining obstacle is trust. Can industry, government, and environmental groups build enough trust to find a more efficient way of getting the same or better results? If industry wants more self-regulation, the first step is better environmental performance — better than required and better than expected. Another step is transparency, with openness and candor. With trust, honest mistakes are not immediately punished.

This book makes a compelling case to proceed with self-regulation. It offers thorough analyses of some of the opportunities and the pitfalls. And it is a compelling read for such a serious topic.

Matthew Arnold
Senior Vice President
and Chief Operating Officer
World Resources Institute

Preface

I believe that we can strike a balance between industrial growth and environmental protection. Throughout my career I have seen blatant disregard for the environment and indifference to environmental regulation by industry. I have also witnessed proactive environmental improvements by firms once thought to be the worst polluters.

Let me give you a brief background about why I think it is time to look at our environmental protection policies a little differently. I began working in the environmental field as an inspector in the New Jersey Department of Environmental Protection during the advent of RCRA. My mindset was that I was going to help clean up the environment, and I would force the polluters to pay for their transgressions. I had responsibility for northern New Jersey, one of the most industrialized parts of the country. During my 4-year tenure at the DEP I saw a lot of disregard for the environment. I investigated plants potentially responsible for the deaths of two sewer workers because of the illegal discharge of solvent waste. I was threatened with a gun by a farmer (yes, there are farms in New Jersey) who had an illegal waste oil facility on his property. And I went on several raids with the Division of Criminal Justice to gather evidence for violations of criminal statutes. When I came to visit facilities unannounced, the atmosphere was adversarial. The company didn't want me there. I was on a mission to find violations. After spending 4 years fighting with industry, I became very frustrated with this casual attitude toward compliance.

So I know how industry can pollute the environment. I also became disenfranchised with the agency's red tape and inefficiencies. It seemed as though we were performing inspections, finding violations, and taking no enforcement action. It took over a year to issue a simple violation notice; layer after layer of agency management had to sign off before issuance, and when it was finally issued it had errors in it. I was wasting my time, getting paid and not accomplishing a thing: and rather than complain about things, I decided to get out. When I was given the choice to go to the EPA or to industry, I thought perhaps I could make more of a difference from inside industry. Up to that point I had planned a future as a career regulator, thinking it would afford me the best opportunity to positively impact our environment.

I was right about having a greater effect from within. In the early and mid 1980s the company I worked for launched an environmental auditing program and initiated internal policies that went beyond regulations — the elimination of underground storage tanks (which is not required by law) and voluntary reductions of waste, water use, and toxic chemicals. Initially it was challenging to get programs going; but as more focus was placed on environmental issues, care for the environment became part of the way we did business. Not only were we concerned with compliance, but we also strove to perform beyond compliance.

This was not only a phenomenon for the company I worked for but for many companies. Companies really want to be in compliance. This is evidenced by widespread attendance at industry group meetings and extensive discussions about interpretation of new and existing regulations. It is evidenced by significant attendance at regulatory training seminars and the fact that environmental auditing programs are a part of most major corporate governance programs. It is evidenced by the voluntary emission reductions and voluntary reporting of environmental performance.

Gone are the days of air pollution episodes, rampant illegal dumping, and the discovery of new Superfund sites. The environmental industry has slowed down. We finally got out of the discovery and study mode and into cleaning up sites.

The current regulations have brought us far. The U.S. has accomplished much: despite a larger population, we have cleaner air and water. The new concerns are more issue-specific and international — global warming, endocrine disrupters, and product take-backs. However, we have reached a plateau. There are no new major regulations on the horizon. If we continue at our current rate, we will only have incremental environmental improvement.

It is clear that the American public wants a clean environment, but no one wants more taxes or more spending on environmental programs. The challenge is how to get continuous improvement for the environment under these circumstances. I believe that at least part of the answer lies in enforcement theory — a certain percentage of firms will comply with the regulations no matter what, and a certain percentage will only comply if they feel that it is in their best interest to comply. How can we encourage the companies in the first category and motivate the second category to voluntarily embrace beyond-compliance initiatives?

Much can be learned by exploring voluntary and self-regulation schemes such as Responsible Care, OSHA's VPP, Project XL, and StarTrack. Many companies have signed onto these voluntary programs and have gone even further with voluntary commitments. Why can't we take advantage of companies that are willing to regulate themselves? Self-regulation means having firms agree to regulate themselves and to go beyond regulations to improve their environmental footprint through voluntary emission reductions and programs such as EPA's Energy Star. It also means that the agency never gives up its right to inspect. The EPA and states should coordinate self-regulatory programs and have the proper checks in place to ensure that there is no backsliding. The key is getting more and more firms interested in joining up and agreeing to regulate themselves. If companies agree to go the extra mile, why not use their greenness to our collective advantage and focus more on those who need more help? I encourage you to take a look at this possibility with me and consider the promise it holds for our future.

Acknowledgments

There are many who played a role in my being able to write this book. Special thanks go to my wife Ronnie, my son Alphonse, and my daughter Marissa for giving me the time to research, study, and write. I know that I was not able to give them all the attention that they deserved during this work.

Thanks to my parents, Alphonse, Sr. and Elvira Iannuzzi, for making it possible for me to further my education and instill in me the desire to learn and the discipline to finish what I start.

Thanks to my entire doctoral committee, who helped me form the foundation for this book — Patricia Heffernan, Dr. Chris Noah, Dr. Joe Meeker, Dr. John Opie, and Dr. Dan Watts. Thank you for your insight and dedication to pedagogy. I especially would like to thank Dr. John Tallmadge of the Union Institute. Though he was a hard taskmaster, he really was a blessing to me.

A very heartfelt thanks to my employer, Johnson & Johnson. I can't imagine a better company to work for. I appreciate all the support that made it possible to accomplish this. I would particularly like to thank the bosses that I had while working on the material for this book — Brenda Davis, Tony Herrmann, Vivian Pai, and Karl Schmidt. Your encouragement allowed me to learn and grow.

I can do all things through Christ which strengtheneth me. Philippians 4:13

Al Iannuzzi

The Author

Alphonse Iannuzzi, Ph.D., CHMM, QEP, is currently employed by Johnson & Johnson as a director in the WorldWide Environmental Affairs group, where he spearheads Johnson & Johnson's design for the environmental program, regulatory outreach initiatives, and remediation. He has extensive experience as an environmental consultant and has gained reality-based experience as a hazardous waste inspector for the New Jersey Department of Environmental Protection. Dr. Iannuzzi earned his Ph.D. degree in Environmental Policy from the Union Institute in Cincinnati, where he researched voluntary compliance and self-regulation. He is a Certified Environmental Management System lead auditor for ISO 14001.

Contents

chapter one

Introduction

Current regulatory model

History of the environmental movement

I was given a gift of an old book entitled *Our Environment: How We Use and Control It*, written in 1934. One section that caught my attention stated, "Man, himself, is perhaps the greatest factor in changing or affecting his environment. He utterly destroyed the great auk, the marsh hen, and the beautiful passenger pigeon" (Wood 1934).

Throughout history we have adversely affected our environment — the list of extinct and endangered species is one shocking indicator of that fact. It seems that we need wake-up calls such as the extinction of various species, grossly contaminated water bodies, or chemical explosions before we take our actions and their consequences seriously. Throughout world history, there have been warning signs to which we did not give the proper attention.

Environmental issues have been of concern in the U.S. for many years. Adopting conservation as a policy was discussed as early as 1908 at a governors' conference on conservation. Considerable debate arose between those who favored preservation and those who argued for *right use* of resources during Theodore Roosevelt's administration. This was an early time in American history to discuss issues so familiar to us as we enter the 21st century (Gottlieb 1993). In the first half of the 20th century, forward thinkers such as John Muir and Aldo Leopold shaped worldwide environmental policy, and many articles and books were published on environmental preservation. Yet, even with early words of caution about environmental issues, the prosperity of American industry yielded many environmental nightmares such as Love Canal, Times Beach, the air pollution episodes of the 1960s, and the Bhopal disaster.

Many consider the publication of Rachel Carson's *Silent Spring* (1962), which highlighted the dangers of pesticides and led to celebration of the first Earth Day in 1970, as the beginning of the environmental movement (Morelli 1999). In response to severe problems such as those expressed by Carson and the resulting public outcry, the U.S. government began passing

1

environmental laws. Air pollution problems brought us the Clean Air Act; uncontrolled hazardous waste sites, such as Love Canal, brought the Super-fund legislation. Although the practice of writing acts and regulations in response to out-of-control situations is analogous to fighting fires (a quick response to an ominous situation), thousands of pages of rules were posted by the Environmental Protection Agency (EPA) in the *Federal Register*.

Environmental regulation has grown exponentially from the first days of the modern environmental movement to the 1990s. Of the 82 major rules written in 1990, 21 of them (25%) were environmental (Fiorino 1995). State and federal environmental enforcement agencies were formed to ensure compliance with these rules. Significant environmental improvement has resulted; we obviously have cleaner air, water, and land than we did in the 1970s.

Shortcomings of the current regulatory system

Many scholars have commented that, although there has been environmental improvement, several shortcomings remain in the current system. For example, the system does not take a holistic approach to the environment; it does not require pollution prevention at the source; and it is very complex, expensive, and burdensome to maintain. The concerns are summed up by John Morelli of the Rochester Institute of Technology, who describes our current approach as an outmoded prescriptive command-and-control system based on problems that existed 25 years ago (Morelli 1999). Robert Repetto, an economist for the World Resources Institute, states that the U.S. spends $200 billion a year on environmental protection — more than any other country. In 1990 this amounted to 2.1% of our gross domestic product. Our regulations require best available control technology without considering cost or impact on the local environment. The Superfund program has spent almost a quarter of a billion dollars without making significant progress toward cleaning up waste sites (Repetto 1995). Not surprisingly, questions have been raised whether we have gone too far and have put ourselves at a competitive disadvantage due to overregulation.

**Frequently Cited Problems with the
U.S. Regulatory System**

- Does not take a holistic view of the environment
- Does not require pollution prevention at the source
- Very complex
- Expensive and burdensome
- Adversarial system

Think tanks and various environmental organizations have published myriad reports critiquing our environmental management program, claiming it is time for a change in direction. Resources for the Future (RFF), a

national public policy research organization, performed a detailed evaluation of the U.S. environmental regulatory scheme. The group reported that
the current system is *so complex and disjointed* that no one person can understand all the requirements. There are hundreds of laws and agencies and
thousands of interest groups. There are nine major federal pollution control
laws dealing primarily with media-specific concerns (Clean Water Act, Clean
Air Act) and hundreds of minor federal laws. The laws are complex and are
getting more detailed with every iteration. Prior to 1970 the Clean Air Act
was only 22 pages long. The 1970 amendments were 38 pages long with
12 deadlines; the 1990 amendments are more than 300 pages long and contain
162 deadlines.

RFF maintains that the laws are complex, unrelated to each other, and
lacking a long-term vision. Many of the laws dictate how the standards will
be met and rely on agency inspections and monetary penalties and jail terms
as incentives to comply. Ironically, the flagship environmental initiative of the
Clinton Administration was Project XL, a program designed to *"circumvent
the inflexibility of existing pollution control laws."* It is a telling sign that the
current system has serious flaws when the major "reinvention" initiative of
our government is to evaluate methods of getting around the current regulations to improve the environment. RFF concluded that, though we have
made many environmental advances, *the system is deeply and fundamentally
flawed* (Davies 1997).

Environmental regulation has also been described as intrusive and filled
with paperwork burdens. According to an Office of Management and
Budget (OMB) estimate, *the EPA imposed 104 million hours* for businesses and
state and local governments to complete reports in 1995. The Chemical
Manufacturers Association analyzed 1994 data and found that the water
pollution control program imposed *three times* the burden of any other EPA
program, requiring *29 million hours of paperwork*. The air program required
9 million hours, and the Resource Conservation and Recovery Act required
7 million hours (Repetto 1995).

Other prominent scholars such as Michael E. Porter, professor of business
administration at the Harvard Business School, and Claus van der Linde,
professor at the International Management Research Institute in St. Gallen,
Switzerland, also see fundamental flaws with the current U.S. regulatory
system. They believe that current regulations *discourage innovation* and risk
taking, and they suggest that industry be given more flexibility to achieve
environmental goals by focusing on the production process rather than on
end-of-pipe control (Porter 1995). Similarly, research by scholars such as
Robert Kagen of the University of California at Berkeley indicates that
adversarial legalism has caused inefficiencies in the current system and suggests
that a more cooperative program is necessary (Kagan 1991).

Despite the weaknesses of current policy, environmental issues are very
important to the American public because it wants clean air, clean water,
and minimal use of natural resources. "An overwhelming majority of voters
think that environmental protection should be maintained or strengthened"

(Repetto 1995). A Gallup poll taken in April 2000 concluded that 96% of the 1004 adults surveyed were "concerned about the environment." A majority (55%) considered environmental problems to be "extremely serious" or "very serious," and only 5% believed that environmental problems were "not serious" (Neville 2000).

Juxtaposed to the desire for a clean and livable environment are strong sentiments to reduce taxes and government spending. Many state environmental agencies have had their budgets cut (BNA 1999). A report critiquing the U.S. environmental regulatory system states, "for the foreseeable future, *neither EPA nor the states will have enough money to implement all of the legally required pollution control functions*" (Davies 1997). Considering these facts, society faces a major problem — how to achieve the public's demand to protect the environment with fewer agency resources.

Because of the many calls for change, pressure has mounted on both federal and state environmental agencies to manage environmental issues differently. Several innovative initiatives have emerged that give industry increased flexibility to comply with environmental regulations and reduce enforcement costs. Such programs, which include the EPA's Project XL and voluntary agreements like the toxic release reduction initiative 33/50, aim for environmental improvements with less regulatory agency oversight and thus less cost to the government. The majority of these programs relies on some form of cooperation between government and industry.

For the foreseeable future, neither EPA nor the states will have enough money to implement all of the legally required pollution control functions.

Compliance and enforcement theory

Before discussing such programs in detail, it is important to understand the purpose of environmental policy. Because environmental regulations are meaningless if not followed, compliance is critical to "realizing the benefits envisioned by environmental policy." All the rules and regulations "put in place to protect public health and the environment amount to empty words and deeds without compliance" (Tietenberg 1992). Consistent enforcement not only ensures the credibility of the regulations but also prevents noncompliant companies from gaining an economic advantage over those that do comply. *Compliance* means that the regulated company is "achieving the required environmental standards, regulations or permit conditions by meeting expected behaviors in processes and practices." *Enforcement* comprises the "legal tool(s) … designed to compel compliance," including the sanctions or penalties that result from noncompliance (Tietenberg 1992).

A commonly held enforcement concept is that of *deterrence*. The precepts of this concept are that a strong enforcement program deters the regulated community from violating. There are four deterrence elements:

- The credible likelihood of detection of the violation
- Swift and sure enforcement response
- Appropriately severe sanction
- Each of these actions is perceived as real (Tietenberg 1992).

A useful perspective when considering enforcement policy is that most federal compliance strategies are formulated with the assumption that most businesses will comply with most of the regulations most of the time (Grumbly 1982). Further, "there will always be *5% of individuals that will violate no matter what, 20% that will comply no matter what, and 75% that will comply only if the violators are punished and/or the requirements are perceived as non-arbitrary*" (Tietenberg 1992). Therefore, an effective enforcement policy should consider that a significant group of companies intend to follow the rules, but a much larger majority needs clear incentives to do so.

What an Enforcement Policy Should Consider

- 5% Will Violate No Matter What
- 20% Will Comply No Matter What
- 75% Will Comply Only if Violators are Punished

Understanding compliance monitoring is also important because this is the method of ensuring conformance to environmental rules. Compliance monitoring includes all the activities by government agencies and the regulated community to gather information to determine the compliance status of a company. Some of the principles of compliance monitoring are that "self-awareness and self-monitoring will guide polluters to take preventive measures to ensure compliance" and a "*credible likelihood of detection* by regulators is necessary for deterrence." It also provides evidence for enforcement actions and data for measuring environmental improvements (Tietenberg 1992).

Current enforcement practice

In order to understand current enforcement practice, it is useful to look at the *Policy Framework*, a 1984 enforcement agreement developed by state and federal officials from all EPA programs. This document defined the expectations, roles, and relationships for an effective enforcement program. The *Policy Framework* has three levels of formal enforcement responses. An initial violation can be addressed by several actions: phone calls, site visits, warning letters, notices of violation, and formal administrative complaints. If problems persist beyond a stipulated date, a formal enforcement response should be made. Formal actions include a lawsuit or an administrative response that is independently enforceable. The second and third levels of response mandate that the violation be corrected and typically include a penalty or other sanction. One of the keys to this policy is a timely and appropriate enforcement response (Tietenberg 1992).

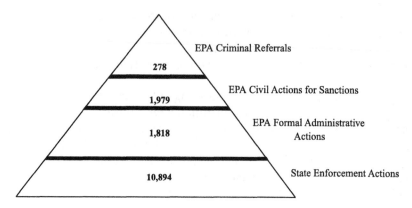

Figure 1.1   Fiscal year 1997 enforcement activities.

Much has been said about ways to improve the current environmental compliance system. An important question to consider is how to determine if an enforcement program is effective and appropriate. The current environmental enforcement system has been criticized because it focuses on the number of penalties issued and criminal cases filed rather than on environmental improvement. Alexander Volokh of the Reason Foundation, a national public policy research organization, concludes that *"new, more effective and fair measures of success must be found"* (Volokh 1996).

To further support this argument, the EPA's 1997 *Enforcement and Compliance Assurance Accomplishments Report* is weighted toward the traditional measures as evidenced by the following statement and table that documents the effectiveness of its activities (EPA, 1998).

> FY97 was another significant year for the EPA in terms of its enforcement program. The Agency reached new heights in the number of referrals (704) to the Department of Justice (DOJ), as well as in fines and penalties ($264.4 million). Of the referrals to DOJ, 426 were civil matters and the remaining 278 were criminal matters. Figure 1.1 presents the total enforcement activities taken by EPA and the states in FY97. As shown, the states accounted for 10,894 enforcement actions (EPA 1998).

To its credit, the EPA is starting to realize that enforcement statistics are not the only means of determining the effectiveness of enforcement and compliance. A press release in January 2001, highlighting the effectiveness of its programs, not only focused on *record* civil and criminal enforcement and *years of criminal sentences* assessed, it also indicated how many pounds of pollutants were reduced as a result of enforcement settlements, compliance assistance given, and the amount of self-disclosed and corrected violations. Nevertheless, the disconcerting term used in the press release is *record*. There

Highlights of FY 2000 EPA Enforcement and Compliance Assurance Activities

Civil and Criminal Enforcement

- A record 6027 cumulative civil, criminal, and administrative enforcement actions
- $224.6 million civil and criminal penalties
- A record 1763 administrative complaints and 3660 administrative compliance orders from FY 1999 (almost double that of FY 1999)
- 146 years of criminal sentences
- $122 million in criminal fines (second highest ever and almost double that of FY 1999)

Pollutants Eliminated

- Enforcement settlements reduced
  - 905.0 million pounds of contaminated soil and sediments
  - 11.6 million pounds of chromium
  - 12.2 million pounds of fecal coliform
  - 116.9 million pounds of solvents
  - 20.8 million pounds of PCB waste
  - 80.0 million pounds of lead
  - 7.4 million pounds of asbestos
  - 1.0 million pounds of ozone layer–depleting CFCs

Compliance Incentives

- Continued to expand the self-disclosure (audit) policy, with 430 companies disclosing potential violations.
- Hot lines, workshops, and guidance materials reached more than 450,000 regulated entities (36% increase from FY 1999).
- Ten Internet-based National Compliance Assistance Centers are available.

Source: Highlights of FY 2000 EPA *Enforcement and Compliance Assurance Data,* January 19, 2001.

were record numbers (6027 cumulative actions) of civil judicial, criminal and administrative enforcement actions (EPA 2001).

Some scholars such as Volokh agree that the reliance on enforcement statistics is not a healthy indicator of enforcement programs. Morelli compares this to whether a police department would be considered to have done a good job if it had more arrests and convictions. He points out that when a police department is doing a good job, there are fewer crimes and fewer convictions. He argues that if the EPA continues to measure its effectiveness using fines and penalties to keep the "bean count" high, it will be pressured to cite companies for less significant violations (Morelli 1999).

These points make a valid argument for not relying so heavily on enforcement numbers. They suggest that we do need to have a strong enforcement program; but fines, penalties, and criminal prosecutions should be a last resort, used only after the agency has tried to encourage compliance or where obvious environmental damage occurred from noncompliance. Is the activity of measuring sanctions synonymous with improving the environment? Many do not believe it is.

Where Is the Focus?

Less crime and fewer convictions are indicators of a police department doing a good job. Why doesn't the EPA take the same tack and focus on environmental improvement rather than "record" numbers of fines and violations?

Complexity of regulatory compliance

Before eagerly issuing fines and violations, we should give some consideration to the complexity of complying with the hundreds of pages of federal and state environmental regulations. Even the former EPA administrator, Carol Browner, has called the current environmental regulatory scheme "a complex and unwieldy system of laws and regulations and increasing conflict and gridlock." In 1993 the *National Law Journal* reported that two thirds of corporate lawyers surveyed indicated that their companies had violated some environmental statute within the past year because of *uncertainty and complexity,* and 70% believe that it is impossible to be in full compliance with all the state and federal environmental laws" (Volokh 1996).

Former EPA Administer Carol Browner

Referred to the current regulatory scheme as a complex and unwieldy system of laws and regulations.

Another indication of the complexity of environmental laws is this statement from a judge regarding the Resource Conservation and Recovery Act: "The people who wrote this ought to go to jail" (Volokh 1996).

Sometimes the permits mandated by the laws are even more complex than the laws themselves. For example, a National Pollutant Discharge Elimination System permit required by the Clean Water Act often runs to 100 pages or more, including effluent limitations, sampling methods, reporting requirements, and recordkeeping. It is not easy for industry to comply with such complex regulations, even when striving for good corporate citizenship. In my professional career, I have spent hundreds of hours reading regulations over and over trying to understand what they mean. A common request by manufacturing sites that I have supported is to interpret the regulations and put them into plain English so they can understand and comply.

Effective environmental policy must therefore take into account the difficulty of compliance. Issuing fines and penalties for failure to comply with regulations that are difficult to understand detracts from the goal of improving environmental quality. An extreme example is the case of a major chemical company in Texas threatened with a lawsuit that could result in fines of millions of dollars because of the EPA's disagreement with the state of Texas interpretation of an air pollution requirement. Ten years after the company had been issued an interpretation by the state of Texas, the EPA disagreed with the state's interpretations and wanted to impose a retroactive fine of $25,000 per day per violation (Volokh 1996). This is an extraordinary case, but it highlights the disconnection with environmental improvement and fines.

Agencies in other states have taken a more nonconfrontational approach. A midwestern firm realized that its already underway construction project had violated the wetlands provisions in the Clean Water Act and disclosed this to the regulatory agency. Rather than seek criminal penalties, the state government worked with the company and permitted the firm to fund the cost of restoring a nearby portion of wetlands. Environmental harm and intentional noncompliance, not "bean counting," should be the factors that drive an enforcement action. The big picture needs to be kept in mind. Environmental quality is the goal, not prosecutions. Awareness is also needed about any regulatory program that forces the collection of fines for its own viability. An example is the South Coast Air Quality Management District Board, which uses fines to help fund its operations (Volokh 1996). This type of program perpetuates adversarial relationships that have been commonplace for years.

Regulatory agency approach to compliance

Regulatory agencies use a variety of methods to encourage compliance. A useful way to understand the compliance options is to look at two divergent mindsets that regulatory agencies can take — deterrence and compliance. *Deterrence* is a strategy that relies on detecting violations and punishing violators. It tends to be accusatory and adversarial and usually results in formal legal processes (Hawkins 1984). Conversely, *compliance* depends on developing a relationship between the regulator and the regulated industry. It is based more on negotiation and rarely relies on the imposition of sanctions. The primary difference between these concepts is that deterrence focuses on finding violations, whereas compliance focuses on achieving the intended purpose of the regulation — environmental protection (Hawkins 1984).

Various Regulatory Agencies' Approaches

- *Deterrence* — Relies on violations and punishment
- *Compliance* — Relies on negotiation and relationship building
- *Amoral Calculator* vs. *Corporate Citizen*

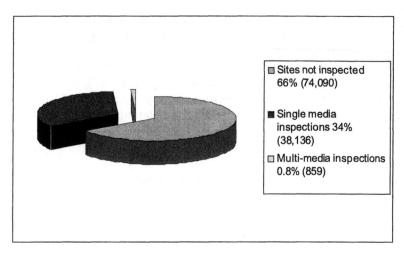

Figure 1.2   EPA inspection rates at large facilities over a 2-year period.

The perception of the regulatory agency is important in determining how to enforce environmental regulation. Kagen uses some terms that help us understand how the agency can shape the enforcement style. One term is *amoral calculator,* where the industry is thought to be consciously devising ways to get around the regulation. The opposing view is termed *corporate citizen*, where the industry is willing to work with the regulator to achieve common goals (Kagan 1983).

Inspection rates

An additional point to consider is that *no enforcement program can detect all the violations all the time because inspection resources are always limited.* Consider the current level of inspections. According to an EPA report, there are approximately 850,000 regulated facilities nationwide, of which 112,226 are considered large sources, defined as one of the following:

- A major air pollutant emitter (Clean Air Act)
- A synthetic minor source (a facility with significant air pollutant emissions, but not enough to be considered a major facility) (Clean Air Act)
- A large-quantity hazardous waste generator under the Resource Conservation and Recovery Act (RCRA)
- A major wastewater discharge source according to the Clean Water Act

In the 2-year period from 1996 to 1998, the EPA reported that 38,136 of the large facilities received a single media inspection (34%). Of these inspections, only 859 facilities, *less than 1%*, received an inspection in all three media (air, water, and hazardous waste). This left 74,090 large facilities that were not visited. Enforcement actions resulting from the inspections occurred at

Table 1.1 Nationwide Inspection/Enforcement
by EPA and State Agencies from 1996 to 1998

	CAA	CWA	RCRA	Total
Large facilities	44,408	6,807	61,011	112,226
Inspected within last 2 years	23,196	5,967	8,973	38,136
	52%	88%	15%	
Enforcement actions	1,252	655	725	2,632
	5%	11%	8%	
Notice of violations	2,706	1,309	3,669	7,684
	12%	22%	41%	

Source: Hale, Rhea, *The National Expansion of StarTrack*, U.S. Environmental
Protection Agency, Boston, 1998, p. 12.

7% of the sites (2632), indicating that a majority of the sites inspected were
not causing serious problems (Hale 1998). The EPA states that the noncompliance data may be understated; nevertheless, these figures suggest that
inspections are performed at a low rate and may not target poor performers.
Also, keep in mind that we have not even considered potential problems at
less informed, smaller facilities (about 738,000 sites).

Coaxing compliance

Some regulatory agencies, such as the Wisconsin Department of Natural
Resources, are rethinking their traditional approaches to enforcement and looking at ways to focus their resources for maximum benefit. Wisconsin came to
believe that the current system targets companies that largely obey environmental laws and began to wonder if it was using the wisest method for spending
public funds. These agencies began to question this "suspicion-driven system"
in which "distrust ... wasted attention on those who don't need it and diverted
attention from unregulated and unsolved environmental problems." In
response, they developed a system that grants flexibility and less oversight to
companies that have good environmental performance records (Meyer 1999).

Newer methods to encourage compliance are noted in the EPA report
entitled *Protecting Your Health and the Environment through Innovative
Approaches to Compliance: Highlights from the Past 5 Years.* In this report the
EPA extols the virtues of encouraging companies to self-audit, disclose nonconformance, and take corrective measures in return for leniency on the part
of regulators. The EPA also speaks about taking holistic approaches such as
multimedia assessments of operations, use of environmental management
systems, and assisting facilities with compliance. Although the EPA still
gauges its own performance by measuring inspections, fines, and violations,
it also admits that such figures "do not help us measure the state of compliance with environmental laws, the environmental results achieved, nor the
degree to which program objectives are being met and noncompliance problems are being addressed" (EPA 1999).

The EPA and some state agencies obviously feel that the use of incentives is also an effective tool for good performance, such as relief from self-reported violations and flexibility. These new approaches show that there are ways of working with the regulated community to persuade compliance.

EPA Admission

The EPA admits that fines and violations do not help us measure the state of compliance with environmental laws, the environmental results achieved, nor the degree to which program objectives are being met and noncompliance problems are being addressed.

I can speak from my own experience as a hazardous waste inspector for the New Jersey Department of Environmental Protection from 1980 to 1984. I was an unwelcome visitor at many facilities and mostly ignored when I pointed out noncompliance with the regulations. I recall being instructed that, for the first round of inspections, we should just issue warnings via written violation notices for noncompliance since the regulation was new. Upon returning to the same facilities a year or two later, we found that the same violations existed. The companies did not take the regulation seriously until they were faced with severe sanctions.

In contrast, roughly 20 years later, I have attended meetings in which environmental managers of major corporations tried to decide whether a tiny sharps container in a medical department met the definition of placement in a secure area. *Most of today's environmental managers want their facilities to be in compliance.* My company trains individuals to treat inspectors with dignity and respect and measures worldwide noncompliance incidents, regardless of the magnitude, to ensure that they are swiftly and permanently corrected. Such experiences have convinced me that many companies are serious about regulatory compliance and are truly trying to be good corporate citizens.

Many companies are serious about regulatory compliance and are truly trying to be good corporate citizens.

Conclusions about compliance and enforcement theory

It is time to consider a move from a system that typically has been based on deterrence policies to a more cooperative system that also includes compliance agreements with corporate citizens. Considering the fact that there are not enough funds available for the EPA and state agencies to enforce environmental regulation, the prospect of a more cooperative compliance program is especially attractive.

Industry voluntary initiatives and government/industry cooperative programs

Many scholars, think tanks, government officials, and industry representatives have criticized the current U.S. environmental policy and enforcement system. Most recommendations for improvement stress cooperative programs and voluntary initiatives on the part of industry. Voluntary initiatives are those that are not driven by regulatory initiatives; "they are voluntary in the sense that governments do not have to order them to be undertaken" (Gibson 1999). Examination of agency- and industry-initiated voluntary programs can assist in the development of strategies for future action.

Cooperative efforts are strongly advocated by John A. Riggs of the Aspen Institute, an international nonprofit educational institution. In his proposal, called the *Alternative Path*, Riggs suggests that companies and communities design tailored approaches to create opportunities for environmental improvement. His work laid the foundation for the EPA's innovative Project XL, which permits companies to propose programs that go beyond compliance in return for flexibility in complying with regulations (Riggs 1999). In support of regulatory flexibility, former EPA administrator Lee Thomas explains that a company is usually much better prepared than a government agency to know the limits of its preventive capabilities and is capable of achieving environmental improvements on its own initiative (Thomas 1992).

The importance of tapping into industry's self-initiated efforts to protect the environment is highlighted in a report from the National Research Council that depicts the current regulatory approach as an incomplete environmental strategy due to its command-and-control methodology. The report concludes that industry-initiated environmental improvement programs should be an important supplement to government regulatory activity (NRC 1997).

Many companies have initiated environmental programs that go well beyond regulatory requirements. Industry-initiated programs can be more efficient and cost-effective than command-and-control regulatory programs. The National Research Council's Committee on Industrial Competitiveness and Environmental Protection studied industry-initiated programs to determine ways to enhance the current industry efforts. There have been many types of industry self-initiated and self-regulated programs that address areas such as management systems, environmental enhancement of products, and process design. An examination of some of these efforts will demonstrate the potential in encouraging companies to willingly make environmental improvements that are not mandated by law.

Industry-initiated programs

First we should understand what motivates companies to start voluntary programs. Some companies go beyond regulatory requirements by voluntarily improving their environmental impact so as to be viewed as good corporate citizens. According to Dirk Schmelzer, professor of economics at

Europa-Universitat, Germany, companies want to be recognized as "socially responsible by customers, employees and neighbors" (Schmelzer 1999). Others, such as Robert Gibson of the University of Waterloo, believe these programs are initiated because of the ease of implementation. Voluntary initiatives are "more flexible and can be adopted more efficiently." If firms can pick the methods that best fit their operations, they will incur fewer expenses and will implement the programs more quickly. The cost savings that result from preventing pollution are also a motivating factor in pro-actively eliminating pollution (Gibson 1999).

3M's 3P Program

3M reduced 159,000 tons of air pollutants, 29,000 tons of water pollutants, 439,000 tons of sludge and solid waste, and 2 billion gallons of wastewater at a sub-stantial cost savings.

Examples of industry-initiated programs include 3M's Pollution Prevention Pays (3P) program and Dow Chemical's Waste Reduction Always Pays (WRAP) program. The 3P program is one of the oldest, encouraging source reduction of pollutants by pollution prevention. 3M claims that the program has been very successful, reducing 159,000 tons of air pollutants, 29,000 tons of water pollutants, 439,000 tons of sludge and solid waste, and 2 billion gallons of wastewater from 1975 to 1994 in the U.S. at a substantial cost savings to the company" (NRC 1997).

Dow initiated its waste reduction program in 1986. WRAP resulted in a companywide culture change that helped Dow see the value of waste and emission reduction. A team of facility personnel and regional environmental managers selects about 20 WRAP projects annually in the U.S. Success is measured by participation in the program and actual reduction of toxic chemical releases that are reported annually to the EPA on the Toxic Release Inventory report (NRC 1997). Similarly, other companies have set voluntary emission reductions such as:

- Du Pont's goals of zero toxic chemical release, 45% reduction of green-house gas emissions, and zero environmental incidents (Du Pont 1998)
- Monsanto's 90% toxic chemical reduction goal (Cairncross 1993)
- Johnson & Johnson's pollution prevention goals to reduce nonhazardous waste by 50%, toxic chemical releases by 90%, energy and packaging by 25%, and hazardous waste by 10% (Johnson & Johnson 1998)

In some cases entire industry groups have volunteered to perform specific actions that are not required by law — i.e., the chemical industry's Responsible Care program, which addresses chemical management, preparedness, and community involvement (ICCA 1996).

Benefit to business is one reason industry would undertake environmental programs without government prodding. According to Ladd Greeno, "most major businesses spend up to three or four percent of sales on environmental, health and safety performance" (Greeno 1996). Because of the magnitude of environmental costs, leading companies have developed metrics to capture the real costs and business benefits of environmental programs. Some forward-looking companies have recognized the business–environment connection and created ways to use environmental issues for business advantage (Greeno 1996).

An example of strategic environmental thinking is S. C. Johnson's decision to use steel aerosol cans. Aerosol cans were not considered recyclable and were stigmatized as a negative environmental product because of their link to CFC propellants. The company worked with local municipalities, the Steel Can Recycling Institute, and regulators to develop recycling markets for steel aerosol cans. The effort to change the image of the cans and develop recycling markets was key to S. C. Johnson's business plan and was developed jointly with business and environmental managers. As a result, the environmental profile of the steel aerosol can was successfully changed, and it gave the company an advantage over competitors.

Another example is provided by Scott Paper, which took the risk of investing its capital to use post-consumer paper in its products. It was able to make recycled content claims that its competitors could not make. Scott's gamble paid off because it was able to influence the regulatory definition of recycled content and force its competitors to also purchase capital equipment. Scott benefited by being first in the market. Similarly, Proctor & Gamble strengthened its hold on the liquid detergent market by being the first to introduce recycled plastic containers and smaller concentrated portions (Greeno 1996).

Government-initiated voluntary programs

Federal voluntary initiatives

Government environmental agencies need to expend resources to enforce regulations through activities such as writing regulations, performing inspections, monitoring emissions, and permitting. By using voluntary agreements, environmental agencies can bring about environmental improvement at a lower cost. They can use minimal resources and still see environmental improvements. Governments have started to "challenge companies to cut emissions instead of requiring them to do so through regulations" (Gibson 1999).

The EPA has seen the advantage in persuading industry to be proactive and to regulate itself. The agency developed a voluntary environmental program, the Audit Policy, that provides industry incentives for disclosing and correcting environmental regulatory violations. The policy waives gravity-based penalties and does not recommend criminal prosecutions for violations voluntarily disclosed. Penalties are also reduced by 75% for promptly disclosed violations. It is the EPA's hope that this will increase compliance with environmental laws and encourage companies to initiate compliance management programs. This policy includes violations discovered by an environmental audit.

The EPA believes that an audit is valuable and reduces a firm's liability. However, there are limitations to the use of the policy. For example, if a company is a chronic violator of environmental laws, the policy cannot be used. The policy also does not apply to violations discovered by routine monitoring, i.e., the sampling required for permits. The basic idea is that a company would be less likely to hold back on reporting a recently discovered compliance problem if it knew that the agency would be willing to work with it and if the severity of the fine would be much less for self-identified issues (Fox 1996).

> Organizations participating in the Energy Star and Green Lights programs saved more that $600 million in their electric bills while significantly reducing nitrogen oxide, carbon dioxide, sulfur oxide, particulates, and toxic metals.

The EPA has initiated a whole host of voluntary pollution reduction programs, including Green Lights, WasteWise, AgStar, Climate Wise, Environmental Leadership, Energy Star, Energy Star Buildings, Environmental Stewardship, and the 33/50 program. The agency embarked on these initiatives because voluntary programs result in "reduced pollution at low or no cost" (NCEE 2001).

The EPA initiated a voluntary industry/government cooperative program in 1994 called the Environmental Leadership Program (ELP). ELP was designed to increase environmental performance, encourage voluntary compliance, and build relationships with the EPA's stakeholders. The program required building trust between the EPA, state agencies, and industry, but it never caught on and has been placed on hold. An interesting position of the EPA in its ELP program was its willingness to treat facilities demonstrating leadership differently by reducing inspections if the facilities had routine audits through independent third-party verifiers (EPA 1998).

Two very successful industry/government partnership programs are the EPA's Green Lights and Energy Star programs. These are voluntary programs in which participants agree to purchase energy-efficient equipment and lighting (Hogan 1996). Voluntary agreements to commit to implementation of energy-efficient practices and submission of periodic progress reports are signed and sent to the agency. The EPA rewards participants with recognition, awards ceremonies, and public service announcements. The results have been very impressive. As of May 1997, over 2300 companies, nonprofit groups, academic institutions, and state and local governments had entered into the Energy Star and Green Lights programs. *More than 6 billion square*

Table 1.2 33/50 Program Target Chemicals

Benzene	Tetrachloroethylene	Cadmium and cadmium compounds
Carbon tetrachloride	Toluene	Chromium and chromium compounds
Chloroform	1,1,1-Trichloroethane	Cyanide compounds
Dichloromethane	Trichloroethylene	Lead and lead compounds
Methyl ethyl ketone	Xylenes	Mercury and mercury compounds
Methy isobutyl ketone	Nickel and nickel compounds	

Source: EPA, *33/50 Program: The Final Record*, Environmental Protection Agency, Washington, D.C., 1999, pp. 1–4.

feet of building space has been made more energy efficient. The environmental benefit from these programs has been the elimination of 10.5 billion pounds of carbon dioxide and significant amounts of nitrogen oxide, sulfur dioxide, particulates, and toxic metals. The organizations that participated also benefited by *saving more than $600 million in their electric bills* (Dolin 1997).

Perhaps one of the most effective voluntary programs initiated by the EPA was the 33/50 program, which targeted 17 toxic chemicals from the Toxic Release Inventory reports that are required to be submitted to the EPA every year. The reports indicate the amount of release of toxic chemicals from each facility to the air, water, and off-site transfers to waste disposal facilities. Companies were asked to voluntarily reduce their emissions 33% by 1992 and 50% by 1995 from a base year of 1988.

According to the EPA's report, *33/50 Program: The Final Record*, program goals were achieved in 1994 — one year ahead of schedule. About 7500 letters

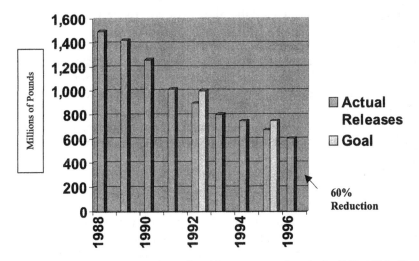

Figure 1.3   Releases and transfers of 33/50 program chemicals 1988–1996. (From EPA, *33/50 Program: The Final Record*, Environmental Protection Agency, Washington, D.C., 1999, pp. 1–4.)

were sent to corporate chief executive officers asking for a commitment, and approximately 1300 companies agreed to participate. The results are impressive: 824 million pounds of toxic chemical releases were removed from the environment. Even better is the fact that progress tracked one year past the program goal to 1996 shows there was a 60% reduction achieved (EPA 1999). The most convincing argument is that it only took sending letters and asking for commitments.

State voluntary initiatives

Innovations have also occurred at the state level, where many believe that cooperative efforts will best occur (Dewitt 1994). Some regulators, such as Kathy Prosser, Commissioner of the Indiana Department of Environmental Management, believe that the states can foster a cooperative, non-confrontational relationship with industry that will lead to environmental improvements. In her opinion state agencies must shift from enforcement and bean-counting to problem solving and measuring environmental results (Prosser 1996).

Other states have been studying the use of self-initiated programs such as management systems. The California EPA has seriously studied the use of ISO 14001 management systems for industry self-regulation and investigating the advantages this voluntary code yields for both government and industry. California has explored ways to encourage industry to participate through regulatory flexibility and relief for enforcement of self-disclosed violations (CalEPA 1996). The Pennsylvania Department of Environmental Protection believes that ISO 14001 will result in cost savings and efficiency and hopes to make the use of ISO 14001 very attractive for companies by reducing inspections and reporting requirements (Begley 1996).

Further state programs are the Massachusetts Toxics Use Reduction Act (TURA) and the California Hot Spots Act. TURA requires users of large quantities of toxic materials to submit annual toxic use reports on their inputs and outputs of toxic materials to the Massachusetts Department of Environmental Protection. The firms must also develop plans to reduce toxic chemicals and waste. Citizens are able to access the data reported by companies. TURA resulted in significant environmental improvement between 1990 and 1997 — 201 million pounds of toxic material reduced, 43.8 million pounds of toxic by-products reduced, and 16.2 million pounds of on-site release of toxic material reduced.

The California Hot Spots Act mandates reporting air toxics emission inventories for certain hazardous substances to local air pollution districts. If the local regulatory agency deems the emissions to be a significant health risk, the reporting company must notify all persons who have been exposed. The emission data and health risk assessments are made available to the public. According to the California Air Resources Board, the Hot Spot inventory information has "increased facilities' awareness of their toxic emissions, leading to ... voluntary reductions of over 1.9 million pounds per year toxics from 21 facilities" (NCEE 2001). Both of these programs suggest that, similar to the 33/50 program, merely making industry results available to the public results in emission reductions.

International voluntary programs

Several notable voluntary initiatives exist outside the U.S. In 1995 the Federal Association of German Industries, along with five other industry associations, initiated a "Joint Declaration of the German Industry on Climate Protection." This is a voluntary agreement to reduce CO_2 emissions or energy consumption by up to 20% from 1990 to 2005. Since the original agreement, an additional four associations have signed on "representing over 71% of industrial energy consumption in Germany and more than 99% of public power generation" (Storey 1999).

One of the most publicized international programs is called the Dutch Covenants. The Dutch government enters into voluntary covenants with industry that require company-specific emission reduction goals in accordance with an environmental business plan. "The business plans require a company to initially assess its ability to reduce emissions, and to continually strive to meet the goals and targets." Failure to meet the covenants will result in civil enforcement and "tightening of its pre-existing license" (Wylynko 1999).

Another Netherlands voluntary program is the Second National Environmental Policy Plan for reduction of greenhouse gas emissions. The nation's target was 3% emission reduction of CO_2 by 2000 compared to 1989 levels. One of the methods used to achieve this target is called Long Term Agreements on Energy (LTAs). These are voluntary agreements with industry to reduce energy consumption that in turn reduces greenhouse gas emissions. The agreements were signed in 1992; as of 1996, approximately 1000 companies participated covering 90% of industrial energy consumption. In return for these agreements, the government would "assure protection from new regulations aimed to reduce energy use and provide technical and financial support to help industry meet the goals" (Storey 1999).

A voluntary agreement in Canada, called the Canadian Industry Program for Energy Conservation, is part of the Industrial Energy Efficiency Initiative of Natural Resources Canada. This program focuses on helping industry to identify energy efficiency barriers and opportunities and helps set energy efficiency targets. The agreement includes 30 industry associations and company groups covering more than 3000 companies and over 85% of secondary industrial energy end use.

Another energy-related program is the New Zealand voluntary agreements with industry to reduce CO_2 emissions. Companies agree to energy reduction targets and sign agreements with the Minister of Energy. The agreements are not enforceable, so there are no penalties for underachievement. As of 1996, 17 voluntary agreements with industry had been signed (Storey 1999).

Industry/environmental group initiatives

Another way to achieve regulatory efficiencies is through industry/environmental group partnerships. A frequently cited example is the Environmental Defense Fund (EDF), McDonald's partnership. In 1990 the EDF was proactive in working with McDonald's to redesign its packaging and eliminate the use of polystyrene foam, thus reducing its waste. The joint environmental group and industry task force reduced packaging waste by 70 to 90% and produced substantial energy savings. This resulted in wins for the environment, industry, and the environmental group (EDF 1991).

Criticism of voluntary programs

Environmental groups tend to be suspicious of cooperative programs. They feel that a lot of hard-fought environmental gains have been achieved, and they see voluntary agreements as a step backward. One of the biggest criticisms is that the agreements are shaped in a noninclusive process "behind closed doors." This lack of access gives an appearance of industry striking "private deals" with government agencies. Environmental groups also feel that the voluntary agreements are used by industry to sidestep regulations and thus lower environmental protection (Muldoon 1999).

Gene Karpinski, executive director of the U.S. Public Interest Group, warns that we must be cautious when implementing regulatory flexibility. He states that we must be flexible but still maintain federal standards and performance objectives. It appears that Karpinski's biggest fear is that recent regulatory flexibility initiatives and movements to have the state as the lead in environmental enforcement will result in a severe weakening of environmental regulation and enforcement (Karpinski 1996). Similarly, Sharon Buccino, an attorney for the National Resources Defense Council, does not believe that

Table 1.3   Examples of Voluntary Programs

Industry Initiatives	Government Initiatives	International Initiatives
• 3M — Pollution Prevention Pays (3P)	• Audit Policy	• Germany — Industry Climate Protection program
• Dow Chemical — Waste Reduction Always Pays (WRAP)	• Environmental Leadership Program	• Netherlands — Dutch Covenants, National Environmental Policy Plan
• Du Pont — Zero toxic chemical release goal	• Green Lights	
• Monsanto — 90% toxic chemical emissions goal	• Energy Star	• Canada — Industry Program for Energy Conservation
• Johnson & Johnson — Pollution Prevention Goals	• 33/50 Toxic Chemical Reduction program	
• Chemical Industry — Responsible Care Program	• MA Toxics Use Reduction	• New Zealand — CO_2 Reduction Agreement
	• CA Hot Spots Act	

Table 1.4 Comparison of Voluntary Initiatives to Traditional Rules

Characteristics	Laws	Voluntary Codes
Cost	High	Low
Developmental process	Difficult: formal, expensive, democratic	Easier: less formal, less expensive, may not be open to all
Requirements and obligations	Complex, expensive to implement, burdensome	Easier to understand, less expensive to implement
Accountability	High: scrutiny by environmental agency inspectors	Lower: depends on reporting requirements
Credibility	High	Lower
Sanctions for noncompliance	Fines and penalties	Usually no sanctions, depends on the program

Source: Kernaghan, Webb and Andrew, M., *Voluntary Approaches in Environmental Policy*, Kluwer Academic Publishers, Boston, 1999, pp. 234, 238.

the current regulatory process needs the radical change many have been calling for and sees the reform attempts as weakening environmental protection (Buccino 1997).

Julian Morris, writing for the Competitive Enterprise Institute, a nonprofit public policy organization dedicated to the principles of free enterprise and limited government, states that ISO 14000 will achieve no environmental improvements but will only help auditors to make money. Morris is very negative about the use of ISO 14000 as an environmental management system and feels that, if mandated by law, it would discriminate against firms that do not have management systems and could become a form of environmental imperialism that would hurt firms in developing nations. He feels ISO 14000 should remain a voluntary initiative and not be considered a replacement for traditional environmental regulation (Morris 1997).

> Corporate America is famous for saying one thing and doing another, so the issue of public trust in self-regulatory programs is significant.

Difficulty in assessing the effectiveness of industry voluntary standards results in distrust of these standards. Nongovernmental organizations (NGOs) believe that businesses need to be more accountable for the claims they make. An example is a study released in 1998 by the U.S. Public Interest Research Group that indicated more than 75% of the chemical industries contacted could not provide basic information about the chemicals they used. This information is a key Responsible Care program objective. Corporate America is famous for saying one thing and doing another, so the issue of public trust in self-regulatory programs is significant (Betts 1998).

Another problem is what is termed a *free ride*. When a group of industries has agreed to a voluntary set of standards, it is possible for some of the poorer performing companies in the industry group not to comply or to give only a token commitment, while the majority of companies are actually conforming to the requirements (Gunningham 1997).

Armin Sandhovel of the German Council of Environmental Advisors in Wiesbaden notes that a difficulty with voluntary agreements is that they "require a great deal of trust in business, because the assumption underlying such agreements is that firms are actually interested in helping to resolve problems." The inference drawn is that voluntary agreements must be transparent, with clear goals. "Credibility and predictability are therefore the most important criteria for assessing voluntary agreements" (Sandhovel 1998).

Conclusions about voluntary initiatives and government/industry cooperative programs

As seen in the above examples, there is a substantial history of voluntary programs. Some industries have embraced such initiatives due to their effectiveness, low cost of administration, and flexibility. State and international agencies have offered alternatives to traditional regulatory solutions by working with new ideas like ISO 14001 and voluntary emission reductions. On the other hand, some environmental groups view such attempts as a way to weaken environmental regulatory enforcement, while others are partnering with industry in voluntary efforts. Regardless of one's opinion of voluntary cooperative agreements, we are about to enter a stage in which more experimentation will occur.

No agency can afford to conduct unlimited inspections ... the challenge is one of priorities and allocation of scarce inspector resources.

Self-regulation theory

Regulatory agencies have a difficult time ensuring that the numerous regulated industries comply with all the volumes of federal and state regulations. As already established, the regulations produce no result if they are not followed. "No agency can afford to conduct unlimited inspections ... *the challenge is one of priorities and allocation of scarce inspector resources*" (Tietenberg 1992). We have seen that, thus far, the allocation of inspectors is incomplete because of limited resources. Inspections are focusing on one program element rather than on multimedia, and the companies that would least likely follow the rules are not targeted. Inspections are important. A study by economists at Duke and Northwestern Universities indicates the importance of targeting enforcement. They noted a statistically significant reduction in effluent wastewater in 4 to 6 months after an inspection with no recidivism in a study of 75 pulp and paper plants from 1977 to 1985 (Tietenberg 1992).

In addition to using inspector resources more effectively, it is important for an enforcement program to develop a complying majority. When the regulatory agency is comfortable that a majority of the companies it regulates are in compliance, it will be able to devote its resources to those that do not comply. One potential way to develop this majority is with an effective self-regulation program. Before more closely examining some innovative self-regulatory programs, it is helpful to see what self-regulation means and understand its history and prospects as a policy instrument.

Definition of self-regulation

Unlike voluntary initiatives, self-regulation is primarily based on compliance with regulations or a set of standards that addresses a perceived gap in regulation. Self-regulation is identified by scholars as "alternate compliance plans conceived by regulated entities, with some degree of government review and involvement" (Steinzor 1998). It enables industry to propose alternative ways to comply with regulations by using different means to achieve environmental goals. Self-regulation programs can also be industry codes of conduct that are sponsored by trade associations designed to improve members' environmental performance. An example is the chemical industry's Responsible Care program (Hemphill 1996).

Neil Gunningham, director of the Australian Center for Environmental Law and professor of law at the Australian National University, distinguishes two types of industry self-regulation — group self-regulation, by which an industry group voluntarily adheres to a code of practice, and individual self-regulation, by which an individual entity regulates itself independent of others. Programs either can be voluntarily initiated by industry or can have government agency oversight/involvement (Gunningham 1997).

Thomas A. Hemphill, of the New Jersey Department of Environmental Protection, states that self-regulation occurs with one of the following:

- "A firm, an industry, or the business community establishes its own standards of behavior when there are no statutory and/or regulatory requirements" or
- A firm, an industry, or the business community establishes "private standards of behavior" that assist in "complying with or exceeding preexisting statutory and/or regulatory requirements" (Hemphill 1993)

In addition to the types of self-regulation already mentioned, there are environmental group-initiated programs like those set up after the 1989 Exxon oil spill in Alaska, the CERES (Valdez) Principles. Environmental groups encourage industries to sign on to their voluntary management principles (Hemphill 1996).

History of self-regulation

Self-regulation is not a phenomenon restricted to the environmental field. It is beneficial to briefly mention some programs in other fields.

1. *Financial futures markets* use self-regulation by a peer group to offset significant opportunities for graft, fraud, and deception.
2. *The Institute of Nuclear Power Operations*, after the Three Mile Island nuclear power plant accident, initiated self-imposed standards. The group ensures conformance to these standards by performing its own inspections and by investigating accidents. This voluntary program has resulted in a significant reduction of nuclear accidents, highlighted by the absence of news stories or adverse publicity for this industry.
3. *The Canadian cable television industry* has a voluntary code that governs its business (Gunningham 1997).
4. *The accounting field* also practices self-regulation. For example, annual financial reports are audited by third-party accounting firms that police themselves (Grant 1996).
5. *In Australia* there are "at least five hundred self-regulatory arrangements administered by industry associations." This practice spans a variety of industries including advertising, professional standards, stock exchange, and futures market controls (Gunningham 1995).
6. *The U.S. Department of Defense* has regulations that require contractors to police themselves, disclose improper conduct, and take proper corrective actions. The contractors must maintain a written code of ethics, perform ethics training for all their employees, maintain a hot line for employees to report improper conduct, and have internal and external audits (Priest 1997).
7. *Many occupations* require a license or certification in order to practice that profession. A self-regulating licensing body that has been given its authority by law issues the license. For example, in Canada a surgeon cannot practice unless he or she is licensed by the College of Physicians and Surgeons. Likewise, an attorney cannot practice unless he or she is a member of the Law Society. A similar system is also in place in the U.S. (Priest 1997).

It is reasonable to conclude that self-regulation can be considered as a potential policy tool for environmental enforcement.

These brief descriptions of programs outside the environmental field indicate that self-regulation is a practice employed in some highly critical industries. There would be severe consequences if self-regulation were not effective in these areas. Therefore, it is reasonable to conclude that self-regulation can be considered as a potential policy tool for environmental enforcement.

For a glimpse of the potential of this policy, it is instructive to look briefly at an early example of self-regulation in the environmental field. In 1990 the EPA became aware of the chemical industry's misinterpretation of reporting requirements in Section 8(e) of the Toxic Substance Control Act (TSCA) via an administrative settlement with Monsanto Corporation. According to Section 8(e), immediate notification is required to the EPA Administrator by any person who manufactures, processes, or distributes a chemical substance or mixture and who has obtained information that reasonably supports the conclusion that such substance or mixture presents a substantial risk of injury to health or the environment. Apparently, companies were performing additional risk assessments of the data received on chemicals and were not immediately reporting this information. As a result of this revelation, the EPA initiated the TSCA Section 8(e) Compliance Audit Program (CAP), which gave companies the incentive of limited penalties for voluntarily auditing their files and reporting studies to the EPA. This was the first program of its kind and resulted in receiving data that would not have otherwise been obtained.

Companies were given a limited time frame to register for the program and had to submit reports within 180 days to the EPA. They also agreed in advance to pay stipulated penalties of up to $1,000,000 per participant. The program was considered successful, with more than 123 companies participating and 90 of these companies having to submit reports. In total, 11,000 studies or reports of adverse effect information were given to the agency that would not have been reported without the program. The EPA had typically received less than 100 Section 8(e) reports per year. As a result of this initiative, new health and environmental information was made available to the public (Jacobs 1997).

In addition, self-certification of compliance by senior company management is prevalent in U.S. laws. The value of this is the attention environmental issues receive from top management after becoming personally liable for false reporting. Self-certification can also include exception reporting of violations. "Source self monitoring and inspections are the most important approaches to monitoring compliance" (Tietenberg 1992). An example is the monitoring required by wastewater discharge permits under the National Pollutant Discharge Elimination System. The use of self-certification is a step toward industry self-regulation. Regulatory agencies seem ready to trust industry to report information necessary to support environmental initiatives.

Reasons for environmental self-regulation

Richard Andrews, professor of environmental policy at the University of North Carolina, describes a "bandwagon of criticism" of current governmental regulation. This includes claims of overregulation, command-and-control regulation, prescriptive regulation that dictates end-of-the-pipe treatment (which moves pollutants from one medium to another), and a lack of resources to perform enforcement (Andrews 1995). Eugene Bardach, professor of public policy at the University of California, seconds many of these claims. He states that "all command-and-control regulation is bound to be excessive; it imposes standardized prescriptions on highly varied problems, and this necessarily produces a large number of cases in which regulatory structures are too costly or are inappropriate to the true situation" (Bardach 1982).

James M. Self, secretary of the Pennsylvania Department of Environmental Protection, in discussing the use of the voluntary standard ISO 14001, indicates that it will "occupy less time and fewer resources" than traditional command-and-control regulation. He envisions the use of ISO 14001 as privatizing environmental protection, easing inspection and reporting requirements for companies that have an environmental management system conforming to the ISO standard (Begley 1996).

One of the most comprehensive and scholarly works on self-regulation was written by Neil Gunningham and Joseph Rees, director of Virginia Tech's Center for Public Administration and Policy. They indicated that one of the driving factors of self-regulation has been regulation overload and counterproductive regulation compounded by reduced resources for enforcement (Gunningham 1997).

Along with these criticisms is a "greening of industry" with the advent of such concepts as design for the environment, industrial ecology, and sustainable development. Some have stated that the greening has occurred because of industry's "enlightened self-interest." It is in a company's self-interest to have a more efficient production process and the ability to avoid fines and liabilities (Andrews 1995). Moreover, some companies are pursuing self-regulatory initiatives like ISO 14001 because they are aware of consumers' concerns for the environment and want to develop their own management systems to cope with the increasing number of environmental regulations. They also see potential competitive advantages and regulatory relief through third-party certification of their environmental programs. These initiatives give companies an opportunity to look holistically at their operations and capitalize on the true regulatory flexibility of programs like Project XL (Begley 1996). Others have turned to self-regulatory initiatives because their policies have come under increasing scrutiny due to expanded and stronger environmental regulation (Hemphill 1993).

Some have initiated self-regulation merely from the threat of additional government regulation — or, as in the case of the Canadian forest industry, market pressure from the European community. Boycotting Canadian products forced the industry into a voluntary set of standards (sustainable forest) to quell market forces. Other industries have initiated self-regulatory programs to ensure their viability — in the wake of the Bhopal disaster in India, the chemical industry initiated the voluntary set of standards called Responsible Care (Gunningham 1997).

Some believe there is a competitive advantage by taking a proactive position and that participating in a self-regulatory program can put a company into a leadership position. The thought is that the laggard or poorly performing companies should be "punished" with more inspections, and proactive corporations should be "rewarded" with more flexibility and the ability to self-regulate (Gunningham 1997). Others believe that "proactive members of the business community may soon be able to opt out of the costly and oftentimes inefficient command-and-control system to become members of a select group allowed self-regulatory power" (Speer 1997). Therefore, if businesses get involved early, they may be able to take advantage of potential benefits sooner.

> Some believe there is a competitive advantage by taking a proactive position and that participating in a self-regulatory program can put a company into a leadership position. The thought is that the laggard or poorly performing companies should be "punished" with more inspections, and proactive corporations should be "rewarded" with more flexibility and the ability to self-regulate.

Self-regulation has numerous potential economic advantages for the private and public sectors. It may result in "faster assimilation of innovative improvements in environmental process and control technologies, streamlining reporting requirements, and paperwork reduction and slowing the growth of public regulatory bureaucracies" (Hemphill 1993). It can reduce the costs for regulators to prepare and review permit applications, renewals and modifications, review monitoring and reporting requirements, and perform inspections (Andrews 1998).

Government sees self-regulation as an opportunity to bring industries into a "normative" behavior. Besides providing lower costs, it can also result in greater speed and flexibility and, since its requirements usually go beyond the letter of the law, it raises standards (Gunningham 1997). Practitioner involvement is thought to result in more effective regulation of the business because it encourages high standards as well as the "opportunity for industry to market itself as environmentally responsible" (Ross 1997).

Evaluation of self-regulation as an enforcement tool is not only a U.S. phenomenon. In "most of the European Union States a debate has begun … about self-regulation" because of the high political and economic costs of government regulation (Sandhovel 1998). Reduced oversight costs are a driver for these programs, especially when the participants are "well-intentioned and well-organized … who would present few problems if left wholly to self-regulate." It is considered by many as a "better, cheaper method of solving the problem than conventional public regulation" (Ogus 1995).

Concerns with self-regulation

In order to present a balanced view of self-regulation as environmental policy, we must study what is written about its limits and problems. The most common concerns were that self-regulation is self-serving, a concoction of industry groups to give the appearance of regulation when none exists, giving the government "an excuse for not doing its job." Self-regulatory standards result in weak standards; "enforcement is ineffective and punishment is secret and mild" (Gunningham 1997).

Some have said that self-regulation is not an alternative to conventional regulation; the best we can hope for is to gain a little efficiency. Also, there is a tendency to have the costs shifted from the regulated industry to other groups. Businesses must follow the *polluter pays* principle: they must bear the full cost of the burden placed on society by their actions (Andrews 1995). Adding another policy instrument like self-regulation, without researching the interaction between environmental goals and the use of more than one instrument, may not be helpful. It is taking an attitude of "more is better" without thinking it through properly (Sandhovel 1998).

Similarly, Rena Steinzor of the University of Maryland School of Law cautions that "self-regulation poses dangers for the environment ... and we should not yield to placating the most vociferous critics." We must be convinced that, at the very least, we "maintain environmental quality and possibly deliver performance superior to status quo." She also states that there needs to be "better incentives for industry and public interest groups to participate in these initiatives" (Steinzor 1998).

Self-regulation has been described as "modern corporatism, where groups have achieved power and are not accountable to 'conventional constitutional channels.'" When the groups regulating themselves are able to make their own rules, it may "constitute an abuse if it lacks democratic legitimacy ... where there is a breach of the separation of powers doctrine." Self-regulatory programs administered by industry groups have been reported to use their power to restrict entry to their group and thereby increase the members' profits. Further, they do not have a good record of enforcing noncompliance among poorly performing members (Ogus 1995).

Apparently *many attempts at self-regulation have failed.* Gunningham wrote that in a vast majority of cases, self-regulatory programs were used more as "an attempt to placate the public and to keep government regulators at bay than as a genuine strategy to achieve broader public interest goals. They are no more than a form of placebo policy designed mainly for its cosmetic effects, a useful way of organizing issues out of politics, a wicked weapon of agenda management" (Gunningham 1995).

An example of a failure at self-policing occurred in the United Kingdom. Under the Environmental Protection Act, Section 34, a *duty of care* provision requires producers of waste to provide a written description of the waste on a transfer note to others that will be managing the waste for them, such as transporters and waste disposal facilities. The regulatory agency relied on self-policing to ensure that this practice was followed. However, in a report about the effectiveness of this program, it was determined that "no transfer note was provided in 70% of cases," and, where a note was provided, "the description of the waste was often vague" (Ross 1997).

A self-regulation program must have accountability built into it. A company commitment to voluntary goals without public scrutiny becomes a public relations ploy. A weakness with programs that are completely self-regulated without any oversight is "corruption" of the system (Iles 1996).

A self-regulation program must have accountability built into it. A company commitment to voluntary goals without public scrutiny becomes a public relations ploy.

Conclusions about self-regulation

Despite the challenges to the use of self-regulation as a policy instrument, the prospect of self-regulation is intriguing. The EPA acknowledges that the "traditional regulatory system, which relies on reporting, inspections, and fines for noncompliance, becomes very cumbersome and expensive to administer when applied to thousands, or even millions, of sources" (NCEE 2001).

EPA administrator Christie Whitman, in a speech to the National Policy Institute in March 2001, indicated that it is time to reevaluate the way the agency does things. She believes that "America is ready to move beyond the command-and-control model ... to a more cooperative scheme where we can come together — stakeholders from every point in the spectrum — to find that common ground." She goes on to say that things have changed — in the more than 30 years that the EPA has existed, the environment is cleaner, and many businesses that once fought against environmental requirements make superior performance part of their business strategy. She is a proponent of new methods to make the agency more effective, like Project XL, where the state that she governed (New Jersey) applied for an XL project for its Silver and Gold Track program. This program provides regulatory flexibility to companies that agree to exceed regulatory requirements (Whitman 2001).

EPA Administrator Christie Whitman

America is ready to move beyond the command-and-control model ... we can come together — stakeholders from every point in the spectrum — to find that common ground.

It seems that the new administration is open to cooperative voluntary compliance initiatives. If a program can be devised that could overcome the pitfalls of self-regulation mentioned above, it would have excellent potential for providing an effective, low-cost alternative to traditional command-and-control enforcement strategy. More detailed consideration of self-regulatory programs can lead to identification of possible effective suggestions and strategies.

Table 1.5 Self-Regulation versus Traditional Regulation

Aspect	Self-Regulation	Traditional Regulation
Rule making	Easier to develop, more flexible and faster to implement, inexpensive	Complex development process, lengthy implementation time, high cost
Agency oversight	Lower administrative resources, more cooperative	High administrative costs, more adversarial
Ease of conformance to standards	Easier to conform, more flexible, less paperwork	More complex, difficult to conform to standards
Public trust	Lower degree of public trust, depends on the amount of government involvement	High degree of public confidence
Stakeholder involvement	Typically low stakeholder involvement, non-inclusive process	More open process, high degree of public involvement
Sanctions for nonconformance	Low or minimal sanctions	High sanctions

Exploring self-regulation and voluntary initiatives

The current environmental system has resulted in much-needed environmental improvement, but we need to move in a different direction to realize environmental improvements under current budgetary constraints. Many proactive industries have initiated environmental programs that go beyond regulatory compliance. They have adopted policies and goals that reach beyond what regulatory agencies require. Government/industry cooperative programs could help address the efficient use of enforcement dollars. It is useful to examine some of the innovative programs initiated by the federal government and industry to see if they result in a more efficient use of taxpayer dollars. If a program can be designed in which industry can police itself with minimal oversight, it could help the government allocate its limited enforcement resources to gain the greatest benefit. The prospect of self-regulation as a policy instrument will become clearer through analysis of industry- and government-initiated case studies in the following chapters.

Bibliography

Andrews, Richard N. L. (1995). Toward the 21st century: planning for the protection of California's environment. *Environment*, 37:25–28.

Andrews, Richard N. L. (1998). Environmental regulation and business self-regulation. *Policy Sci.*, 31(3):188.

Bardach, Eugene. (1982). Self-regulation and regulatory paperwork, in *Social Regulation: Strategies for Reform*. Bardach, E., Kagan, R. A., Eds., New Brunswick, NJ, Transaction Books, p. 328.

Begley, Ronald. (1996). ISO 14000: a step toward industry self-regulation? *Environ. Sci. Technol.*, 30(7298):301, 302.

Betts, Kelly S. (1998). UN commission calls for review of voluntary environmental initiatives. *Environ. Sci. Technol.*, 32(13):303.

BNA (1999). EPA hiring freeze announced as funding cut looms. *BNA Environment Reporter*, 6-18-99, p. 321.

Buccino, Sharon (1997). Environment and the 105th congress: to be or not to be green, Available: http://www.nrdc.org/nrdc/nrdcpro/analys/sbgw0197.html, August 25, 1999, p. 1.

Cairncross, F. (1993). *Costing the Earth: The Challenge for Governments, the Opportunities for Business.* Boston, Harvard Business School Press, p. 257.

CalEPA (1996). *ISO 14000 Pilot Project.* Sacramento. December 13, p. 1.

Carson, Rachel (1962). *Silent Spring.* Boston, Houghton-Mifflin.

Davies, Clarence J. and Mazurek, J. (1997). *Regulating pollution: does the U.S. system work?* Resources for the Future. Available: http://www.rff.org/research/reports/pollrep.htm. April 25, pp. 2, 7.

Dewitt, John (1994). *Civic Environmentalism: Alternatives to Regulation in States and Communities.* Washington, D.C., Congressional Quarterly Press, p. 341.

Dolin, J. (1997). EPA's energy efficiency programs, *Environment*, 39:12–13.

Du Pont (1998). *Sustainable Growth 1998 Progress Report.* Wilmington, DE, Du Pont, p. 1.

EDF (1991). McDonald's corporation — environmental defense fund waste reduction task force. Available: http://www.edf.org/pubs/reports/mcdfinreport.html, pp. 1–4.

EPA (1998). *Enforcement and Compliance Assurance Accomplishments Report: FY 1997.* Washington D.C., U.S. Environmental Protection Agency, pp. 2-1, 2-2.

EPA (1998). Home page — environmental leadership program. Available: http://www.es.epa.gov/elp/frmwk.html, June 8, p. 1.

EPA (1999). *Protecting Your Health & the Environment through Innovative Approaches to Compliance: Highlights from the Past 5 Years.* Washington, D.C., U.S. Environmental Protection Agency, pp. 14, 17, 21.

EPA (1999). *33/50 Program: The Final Record.* Washington D.C., U.S. Environmental Protection Agency, pp. 1–4.

EPA (2001). *EPA Releases FY 2000 Enforcement and Compliance Assurance Data.* Washington, D.C., U.S. Environmental Protection Agency Communications, Education, and Media Relations (1703), January 19, pp. 1–2.

Fiorino, Daniel J. (1995). *Making Environmental Policy.* Los Angeles, University of California Press, p. 72.

Fox, Catherine A. (1996). EPA's new voluntary self-policing and self-disclosure policy. *Remediation*, Spring, pp. 123–126.

Gibson, R. B. (1999). *Voluntary Initiatives: The New Politics of Corporate Greening.* Peterborough, Broadview Press Ltd., pp. 3–5.

Gottlieb, Robert (1993). *Forcing the Spring: The Transformation of the American Environmental Movement*, Washington, D.C., Island Press, p. 24.

Grant, Julia, Bricker, R., and Shiptsova, R. (1996). Audit quality and professional self-regulation: a social dilemma prospective and laboratory investigation. *Auditing*, 15(1):142.

Greeno, Ladd J. (1996). *Environmental Strategy.* Cambridge, MA, Arthur D. Little, pp. 18, 29, 34.

Grumbly, T. P. (1982). Self-regulation: private vice and public virtue. *Social Regulation: Strategies for Reform.* Bardach, E. and Kagan, R. A., Eds., New Brunswick, NJ, Transaction Books, p. 102.

Gunningham, Neil (1995). Environment, self-regulation, and the chemical industry: assessing responsible care. *Law Policy,* 17:58, 91.

Gunningham, Neil and Rees, Joseph (1997). Industry self-regulation: an institutional perspective. *Law Policy,* 19(4363):363–364, 370, 396.

Hale, Rhea (1998). *The National Expansion of StarTrack.* Boston, U.S. Environmental Protection Agency, Region I New England, pp. 10–12.

Hawkins, Keith and Thomas, J. M. (1984). *Enforcing Regulation.* Boston, Kluwer Nijhoff Publishing, pp. 13–14.

Hemphill, Thomas A. (1993). Corporate environmentalism and self-regulation: keeping enforcement agencies at bay. *J. Environ. Regul.,* 3(2):145, 149, 152.

Hemphill, Thomas A. (1996). The new era of business regulation. *Bus. Horizons,* 39(4):26–31.

Hogan, Kathleen (1996). Commentary. *Environment,* 38(7):3.

ICCA (1996). *Responsible Care Status Report.* Brussels, International Council of Chemical Associations, p. 4.

Iles, Alastair T. (1996). Hybrid regulatory systems. *Environment,* 38(7):4.

Jacobs, Jon and Ahearn, Caroline (1997). EPA's enforcement compliance audit programs: increased environmental compliance and reduced civil penalties. *J. Environ. Law Pract.,* 4(6):5, 9.

Johnson & Johnson (1998). Johnson & Johnson environmental health and safety. *Johnson & Johnson Inc. Annual EHS Report,* p. 28.

Kagan, Robert A. (1983). On regulatory inspectorates and police. *Enforcing Regulation.* Hawkins, K. and Thomas, J. M., Eds., Boston, Kluwer-Nijhoff Publishing, pp. 74–77.

Kagan, Robert A. (1991). Adversarial legalism and American government. *Public Policy & Management,* reprinted in *The New Politics of Public Policy.* Johns Hopkins Press, Baltimore, 1995, p. 369.

Karpinski, Gene (1996). Environmentalist fears state flexibility may become rollback. *State Environ. Monitor,* 1(3):20–21.

Kernaghan, Webb and Andrew, M. (1999). Voluntary approaches, the environment and the law: a Canadian perspective. *Voluntary Approaches in Environmental Policy.* Carraro, C., Ed., Boston, Kluwer Academic Publishers, pp. 234, 238.

Meyer, George E. (1999). *A Green Tier for Greater Environmental Protection.* Madison, WI, Department of Natural Resources, p. 5.

Morelli, John (1999). *Voluntary Environmental Management.* Boca Raton, Lewis Publishers, pp. 9, 48, 55, 125.

Morris, Julian (1997). ISO 14000: Environmental Regulation by Any Other Name? Available: http://www.cei.org/MonoReader.asp?,ID=121, August 25, 1999, p. 1.

Muldoon, Paul and Ramani, N. (1999). Beyond command and control: a new environmental regulatory strategy links voluntarism and government initiative. in *Voluntary Initiatives: The New Politics of Corporate Greening.* Gibson, R. B., Ed., Peterborough, Ontario, Broadview Press Ltd., pp. 57–59.

NCEE (2001). *The United States Experience with Economic Incentives for Protecting the Environment.* Washington, D.C., National Center for Environmental Economics. January, pp. iii, 159–160, 164, 173.

NRC (1997). *Fostering Industry Initiated Environmental Protection Efforts.* Washington D.C., National Academy Press, pp. ix, 33.

Neville, A. (2000). A greener battlefield. *Environ. Prot.*, II(10):6.

Ogus, Anthony (1995). Rethinking self-regulation. *Oxford J. Leg. Stud.*, 15(1):97–99, 102.

OSHA (1998). Home page. Available: http//www/osha.govoshprogs/vpp/overview. html, June 4.

Porter, Michael E. and Van der Linde, Claas (1995). Green and competitive: ending the stalemate. *Harvard Bus. Rev.*, September-October, pp. 120–134.

Priest, Margot D. (1997). The scope and limits of self-regulation: an analytic framework and case studies. *Law School.* Ontario, York University, pp. 26–27, 33–34.

Prosser, Kathy (1996). From punishment to problem solving: the new environmental enforcement. *State Environ. Monitor*, 1(3):17–19.

Repetto, Robert C. (1995). *Jobs, Competitiveness, and Environmental Regulation: What Are the Real Issues?* Washington, D.C., World Resources Institute, pp. 1, 17, 29.

Riggs, John A. (1999). The alternative path. The Aspen Institute. Available: http://www. aspeninst.org/dir/polpro/EEE/Alternate.Path/Chapter3.html, August 25, p. 1.

Ross, Andrea and Rowan-Robinson, J. (1997). It's good to talk! environmental information and the greening of industry. *J. Environ. Plan. Manage.*, 40(1):115, 116.

Sandhovel, Armin (1998). What can be achieved using instruments of self-regulation in environmental policy making? *Eur. Environ. Law Rev.*, 7(3):83, 84.

Sayre, Don (1996). *Inside ISO 14000: The Competitive Advantage of Environmental Management.* Boca Raton, FL, St. Lucie Press, p. 232.

Schmelzer, D. (1999). Voluntary agreements in environmental policy: negotiating emission reductions. in *Voluntary Approaches in Environmental Policy.* Carraro, C., Ed., Boston, Kluwer Academic Publishers, p. 56.

Speer, Lawrence J. (1997). From command and control to self-regulation: the role of environmental management systems. *Int. Environ. Rep.*, March 5, pp. 227–229.

Steinzor, Rena I. (1998). Reinventing environmental regulation: the dangerous journey from command to self-control. *Harv. Environ. Law Rev.*, 22(1103):104, 200–202.

Storey, M., Boyd, Gale, and Dowd, J. (1999). Voluntary agreements with industry. *Voluntary Approaches in Environmental Protection.* Carraro, C., Ed., Boston, Kluwer Academic Publishers, pp. 195–199.

Thomas, Lee M. (1992). The business community and the environment: an important partnership. *Bus. Horizons*, 35:4.

Tietenberg, Tom (1992). *Innovation in Environmental Policy.* Brookfield, Edward Elgar Publishing, pp. 22–25, 34–38.

Volokh, Alexander and Marzulla, Roger (1996). *Environmental Enforcement: In Search of Both Effectiveness and Fairness.* Los Angeles, Reason Foundation, pp. 1, 3, 5–8.

Whitman, C. (2001). Remarks of Christie Whitman, administrator of the U.S. Environmental Protection Agency, at the National Environmental Policy Institute, Washington, D.C. March 8, 2001. Available: http://www.epa.gov/epahome/ speeches_031301.htm, April 2, pp. 1–2.

Wood, G. and Harry, C. (1934). *Our Environment: How We Use and Control It.* New York, Allyn & Bacon, pp. 19–20.

Wylynko, B. D. (1999). Beyond command and control: a new environmental regulatory strategy links voluntarism and government initiative. in *Voluntary Initiatives: The New Politics of Corporate Greening.* R. B. Gibson, Ed., Peterborough, Ontario, Broadview Press, p. 169.

Yin, R. K. (1989). *Case Study Research: Design and Methods.* Newbury Park, CA, Sage, p. 11.

chapter two

Responsible Care case study

Background

The first program to examine is Responsible Care because it is the oldest environmental self-regulatory program and is the only one initiated by a group of companies in the same sector. Starting with this case first will result in a unique base to help understand the potential of self-regulation. Our study will cover the history of the program, eligibility for participation, program elements, and an examination of the benefits and criticisms of the program. Finally, conclusions will be drawn about what can be learned about Responsible Care as a self-regulatory initiative.

History and eligibility

In 1984 at least 2000 people died when 20 tons of deadly methyl isocyanate was released from a Union Carbide facility in Bhopal, India. A 1990 opinion poll in the U.S. indicated that the public acceptability rating for the chemical industry had dropped 25%. "Over 60% of the public rated the chemical industry as very harmful to the environment." Only the tobacco industry was rated lower than the chemical industry (Gunningham 1995).

> Responsible Care is the chemical industry's response to the public distrust of its operations that resulted from numerous accidents, the most prominent of which was Bhopal.

Responsible Care is the chemical industry's response to the public distrust of its operations that resulted from numerous accidents, the most prominent of which was Bhopal. It is a voluntary commitment to principles designed to achieve "continuous improvement in all aspects of health, safety and environment (HS&E)." The program began in 1985 when the Canadian Chemical Producers Association launched it as a framework for improvement. It is now coordinated worldwide by the International Council of

Chemical Associations (ICCA). However, each country runs its own Responsible Care program through its nation's primary chemical association. For example, in the U.S. the program is run by the American Chemistry Council (ACC), formally called the Chemical Manufacturers Association (CMA). In the United Kingdom it is run by the Chemical Industries Association. It is through the national associations that the members become signatories to the Responsible Care principles (ICCA 1999).

Program elements

The Responsible Care program consists of eight fundamental features:

- Guiding principles
- Name and logo
- Codes, guides
- Indicators
- Communications
- Sharing
- Encouragement
- Verification

Each country can adopt different criteria for the fundamental features, but they must be similar in scope. These elements will become clearer by examining each one and considering what the U.S. association, the ACC, has put into place as examples. The ACC initiated the U.S. program in 1988. The *guiding principles* are those that have been agreed upon by the member companies and which their CEOs must sign. The ACC lists these as Community Awareness and Emergency Response, Pollution Prevention, Process Safety, Distribution, Employee Health and Safety, and Product Stewardship.

Each association must adopt a *logo* to be used solely by program participants to promote the commitment to Responsible Care. A set of *codes*, guidance, or checklists must be made available to assist companies in implementing the Responsible Care principles in order to achieve better HS&E results. Training on the guidance and intercompany assistance to help achieve program goals is also required.

Indicators for judging program effectiveness must be developed. The ACC has chosen its indicators to be Toxics Release Inventory Emissions, Occupational Injury and Illness Reports, Process Safety Facility Incidents, Transportation Incidents, Employee and Emergency Responder Surveys, and Customer Surveys on Product Stewardship. *Communication* with interested parties is accomplished through a national public advisory panel made up of external stakeholders and more than 518 local community advisory panels in areas where facilities are located (ICCA 2000). The national panel provides input on Responsible Care to the ACC. The local advisory panels interface directly with the neighborhood chemical plant. An annual status report is made available to the public as well as other explanatory materials on Responsible Care.

Table 2.1   Responsible Care Code Performance Measures

Code	Measure
Community awareness and emergency response	Survey of plant community residents, emergency responders near plant sites, and employees Number of community advisory panels (CAPS)
Distribution	Assessment of transportation incidents involving hazardous materials based on DOT databases and Association of American Railroads data
Employee health and safety	Assessment of employee injury data based on OSHA recordable injuries
Pollution prevention	Assessment of toxic release inventory data based on EPA required reporting
Process safety	Collection of annual incident reports by ACC for all relevant process safety incidents
Product stewardship	ACC Measurement of code performance by ACC with a customer perception survey

Source: CMA, *Responsible Care Code Performance Measures,* http://www.cmahq.com/responsiblecare.nsf/a46d0aad0715960a852567230067bd7?OpenDocument, June 1999.

A host of activities occur to accomplish *sharing* and *encouragement* such as annual conferences for Responsible Care members, semiannual meetings of regional executive leadership groups for presidents and CEOs, and Responsible Care coordinator network meetings four to six times a year. In order to assure *verification,* the ACC has had over 140 companies undergo a voluntary management systems evaluation. The review teams consist of industry peers and members of the public, i.e., environmental groups.

Table 2.1, which indicates the code requirements and measurements for ACC members in the U.S., is helpful in understanding the program elements.

It is interesting to note that participation in Responsible Care has become one criterion for membership in ACC. This is an obvious sign of how important the chemical industry considers this program. It feels so strongly about it that failure to participate can lead to expulsion from the ACC (ICCA 1999).

The breadth of coverage of this program is quite impressive. The 2000 Responsible Care Status Report asserts that 46 countries have adopted the voluntary standard, including such nations as India, Malaysia, Thailand, and Zimbabwe (ICCA 2000). In 1998 the members were responsible for 87% of the worldwide chemical production volume (ICCA 1999).

Analysis of Responsible Care

A) Benefits of Responsible Care

With this background on Responsible Care, we can now explore what has been said about its effectiveness. First, what does the ICCA say about the program's success?

The ICCA receives annual progress reports from its members based on the eight fundamental features. Individual member companies perform self-assessments of their programs and report the results to their national associations. Each national association in turn sends its country's data to the ICCA, reporting its progress on a scale that ranges from A (initiating) to D (in place).

In their annual reports, country organizations are evaluated by placement into either of two groups — a group of *mature* programs, consisting of countries that began 8 years ago or more, and all others. Some of the countries in the mature group include Canada, the U.S., and the U.K. As expected, these countries are much farther along in their progress, having a majority of the fundamental features at or approaching the D level. Worldwide, 58% of the eight fundamental features are at the highest level, D. This is an improvement from 47% in 1998. The weakest areas are in *encouragement* and *verification*. Seventy-seven percent of the elements implemented for all countries (level C) compares to 83% at level C for mature countries (ICCA 2000).

Individual member companies perform self-assessments of their programs and report the results to their national associations.

Other accomplishments resulting from Responsible Care include the continuing addition of voluntary initiatives such as a global effort to test and prepare data sets and hazard assessments on High Production Volume (HPV) chemicals. Western European chloralkali producers also voluntarily agreed to six binding commitments to reduce mercury impacts, called the *Madrid Commitments*. Individual country projects from the ICCA Responsible Care Status Report include sponsoring household chemical collection days and participation in various country-specific voluntary agreements to reduce chemical emissions like volatile organic compounds and greenhouse gas emissions. Overall, ICCA claims that significant progress has been made since the last (and first) status report published in 1996. Illustrations of progress cited are increased support given to member companies by the association to improve conformance to the program and the improvement of each country's achievement of program goals (ICCA 2000).

In Europe, companies have reported on their collective environmental improvement efforts for various parameters like SO_2, NOx, and CO_2. Using 1996 as a base year, all parameters have had a decreasing trend. For example, SO_2 emissions have fallen about 30% from 1996 to 1999 (ICCA 2000).

A very positive aspect of the program is the active attempt to engage stakeholders. The 2000 ICCA Responsible Care Status Report documents conversations with stakeholders such as Elizabeth Salter, head of the European Toxics Program for the World Wildlife Fund (WWF), Vic Shantora, director general, Toxics Pollution Prevention Directorate, Department of the Environment, Canada, and Dr. Gatot Ibnusantosa, director general of

Chemical, Agricultural and Forestry-based Industries with Indonesia's Ministry and Trade. To the industry's credit, some significant critiques were printed in its reports. For instance, Elizabeth Salter criticized chemical companies for not being open and transparent. She gives her opinion that the industry has not "moved forward very far yet in terms of communicating and addressing many of the concerns of the public." She also states that, in addressing life-cycle analysis of chemicals, chemical firms have created roadblocks against attempts by the European Commission to publish lists of endocrine-disrupting chemicals. She does compliment the ICCA for its "don't trust us, track us" approach, but goes on to say that some kind of independent body to evaluate performance should be established (ICCA 2000).

Regardless of your opinion of Responsible Care, having traditional adversaries' criticisms in its own document is taking a very big step toward open and honest communication. It is my opinion that this kind of bold step is a real indicator that the industry group is serious about improving its performance, and it probably says more than any of the other data presented.

> Regardless of your opinion of Responsible Care, having traditional adversaries' criticisms in its own document is taking a very big step toward open and honest communication.

In the U.S. the American Chemistry Council, formally called the CMA, has a Responsible Care vision of "no accidents, injuries or harm to the environment" (ACC 2000). The association claims several improvements resulting from the Responsible Care program:

1. Toxic chemical emissions are down 58% from 1988 to 1997, while member companies' production increased by 18%.
2. A 464% increase in the number of community advisory panels was formed from 1991 to 1996.
3. Process safety incidents went down from 554 in 1996 to 512 in 1998.
4. Occupational injury and illness rates decreased from 2.93 to 2.13 per 200,000 hours from 1993 to 1999.
5. From 1992 to 1998, energy efficiency improved 13.5% (ACC 2000).
6. Air, highway, and rail incidents reduced by 776 from 1996 to 1997; and rail incidents per 200,000 hours dropped from 1.68 in 1995 to 1.61 in 1996 (CMA 1999).

One of the most interesting aspects of Responsible Care is the community advisory panel (CAP). The CAPs are designed to build trust and improve dialogue with the communities in which chemical plants operate. The CAP consists of members of the community who meet with plant personnel to discuss issues of common interest. The code also requires annual emergency

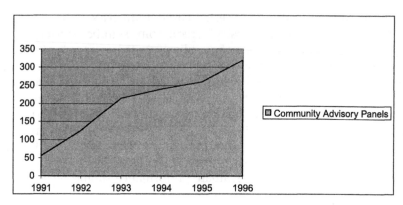

Figure 2.1	Community advisory panel growth in the U.S.

drills to be conducted with the community. The growth of the CAPs over time is a good indicator of the adherence to the code principles. CAPs have increased fivefold in the U.S. since 1991.

> One of the most interesting aspects of Responsible Care is the community advisory panel (CAP). The CAPs are designed to build trust and improve dialogue with the communities in which chemical plants operate.

In the U.S. the involvement of the community goes even farther. Member companies are required to develop performance goals in conjunction with the community, and they are required to report back their results to the community. In addition, as part of the Management Systems Verification (MSV), an audit of a company's adherence to the code — members of the public can participate. In 1999, 29 verification audits occurred. For the first time, a member of the EPA participated in an MSV visit at a Dow Chemical site. To date, 77 ACC members have completed MSV evaluations. An impressive 237 citizens have participated in MSV assessments since its beginning in 1994, including members of CAPs, teachers, clergy, plant neighbors, elected officials, and students (ACC 2000).

The ACC hasn't been reticent about printing unfavorable news on its program either. In its public opinion and employee research, it determined that only 66% of Responsible Care industry employees said they were aware enough of the code to be "familiar" with it. The public opinion polls were not encouraging either; only 20% of Americans polled were aware of Responsible Care — in general, people felt they did not know enough about the industry. However, 84% responded positively when asked if they have a favorable opinion of Responsible Care after it was explained to them. I think

Table 2.2 shows the alignment of the program elements, its measures and results.

Table 2.2   Responsible Care Code Performance Measures

Code	Measure	Results
Community awareness and emergency response	Survey of plant community residents, emergency responders near plant sites, and employees Number of CAPS	Percentage who agree "the chemical industry protects the health and safety of people in the plant communities" Employees — 60% LEPC members — 38% Plant Communities 16% ACC members formed over 300 CAPs, a 464% increase in 5 years
Distribution	Assessment of transportation incidents involving hazardous materials based on DOT databases, and Association of American Railroads data	ACC members reduced their total air, highway, and rail incidents by 776 or 15.6% from 1996 to 1997. Rail incidents per 1000 rail originations dropped from 1.68 in 1995 to 1.61 in 1996 (latest data)
Employee health and safety	Assessment of employee injury data based on OSHA recordable injuries	ACC members reduced injury rates steadily since 1993. Hourly injury and illness rates fell from 2.93 to 2.01 per 200,000 exposure hours from 1993 to 1997
Pollution prevention	Assessment of toxic release inventory data based on EPA required reporting	ACC members reduced releases, transfers, and underground injection of toxic chemicals by 58% from 1988 to 1997 while production increased 18%
Process safety	ACC collects annual incident reports for all relevant process safety incidents	Incidents reduced from 554 in 1996 to 512 in 1998 (a 9% decline)
Product stewardship	ACC measures code performance with a customer perception survey	First-year results indicate that suppliers are doing a good job with traditional product stewardship activities, like providing MSDSs and labels. Work should continue with counseling customers on the safe use of their products and understanding customer product uses (1997 data)

Source: CMA, *Responsible Care Code Performance Measures*, http://www.cmahq.com/responsiblecare. nsf/ June 1999, pp. 1–2.

the chemical industry did itself disservice with the public opinion part of this poll (ACC 2000). I personally would not have heard of Responsible Care if I were not in the environmental protection business. Similar to the international association, the ACC should be commended for its openness in putting seemingly bad news in its annual report. This diffuses the argument that its reports are simply public relations documents.

Others have also noted the positive attributes of Responsible Care as well. Neil Gunningham performed a comprehensive evaluation of Responsible Care, concluding that the program has "considerable virtues" and is the "most sophisticated and advanced self-regulatory scheme yet developed." He rightly points out that there are other industry groups with just as much environmental impact that have taken no voluntary initiative. Although Gunningham is critical of pure self-regulation as a concept, because of many failed attempts as discussed in Chapter 1, he compliments the program by declaring that it is "detailed and far reaching." The attempt to involve the community is a significant step. The program also has the primary benefits of self-regulation, flexibility, and lower costs (Gunningham 1995).

> Responsible Care is the most sophisticated and advanced self-regulatory scheme yet developed.

Philip Abelson, deputy editor for *Science* magazine, mentions several positive attributes of Responsible Care. Among these are taking steps to reduce accidents and reducing the amounts of hazardous chemicals stored on participants' sites. An example is replacing chlorine gas containers with "just-in-time delivery of the chemical" (Abelson 1992).

Another significant component to Responsible Care is public involvement and interaction. National community advisory panels (NCAPs) consisting of "community representatives and independent technical specialists" are formed to receive independent input and critique. Some of the members, such as environmental groups, are traditional adversaries of the chemical industry. The members play a significant role in drafting the codes of practice and evaluating the program. "Industry has gone to great lengths to understand and respond to NCAP concerns." Although the input of the NCAPs cannot veto the actions of the national chemical association, the industry has yet to reject any of their concerns (Gunningham 1995).

In the U.S., local CAPs meet with plant management monthly or bimonthly. Their recommendations are taken seriously and result in action (Abelson 1992). Ten years into the program, 300 CAPs have been formed. Portia Krebs, CMA's community awareness and emergency response coordinator, declares: "we've done much more than we ever imagined we would." Before Responsible Care, "no chemical company would have thought of sitting down with its community and asking what was wrong" (Mullin 1998).

According to Jennifer Howard, Jennifer Nash, and John Ehrenfeld of the Massachusetts Institute of Technology, Responsible Care companies have taken many positive steps. An MIT study interviewed plant managers, Responsible Care coordinators, and community affairs coordinators from 16 medium-sized chemical companies in the U.S., all CMA members. The CMA adopted Responsible Care in 1988; therefore, at the time of the survey, the companies had had about a decade of experience with it. The study determined that one of the most significant achievements was the involvement of employees and the community. "Ninety percent of the plant managers interviewed said that Responsible Care had resulted in a significant or complete change in their interaction with their company's community relations function." Fifteen companies had "created Community Advisory Panels" (Howard 1999).

Another significant improvement is in the area of distribution and transportation safety. Much more is required of chemical distributors than before implementation of Responsible Care. All companies have put audit systems in place to "assess their carrier's safety and handling procedures." The audits have become so common that the transporters have created their own codes of practice. Some companies found that the code provided an incentive to install "new systems to measure and track performance" of their own safety systems (Howard 1999).

Chemical Week's editor-in-chief, David Hunter, made further claims of environmental improvement in an evaluation of the Responsible Care program. He claims that the voluntary standard has prevented accidents, citing as evidence the "two most lethal U.S. chemical industry accidents of the past 18 months" were from companies that did not embrace Responsible Care. The accidents occurred at Concept Sciences in Allentown, PA, where five people died, and at Sierra Chemical in Reno, NV, where four people were lost (Hunter 1999).

Another surprising benefit of Responsible Care is peer pressure. As chemical associations collect and report data, the members "will be subject to greater performance expectations from their peers and outside stakeholders." This expectation will result in better performance (Druckrey 1998).

Jeffery Rayport and Professor George Lodge of the Harvard Business School prepared an insightful case study that gives an in-depth look at Responsible Care through the chemical industry's eyes. The study reveals that if nothing else has been accomplished, the program has received the attention of chemical industry executives and put much-needed focus on their environmental programs. An example is Anthony Sinibaldi, senior vice president for a small chemical company, Standard Chlorine of Delaware, Inc. His company has only two plants and employs 250 people, so at first he was not sure about taking on Responsible Care. However, after pondering the effects of the Chernobyl nuclear reactor accident in 1986, he changed his mind. He once thought that what went on inside his fence line was his company's business. However, the devastation wrought by Chernobyl made him think again and support Responsible Care. Since his company was small, he did not see an advantage for taking on this initiative. Nevertheless, all he had to do was think about having an accident and the headlines reading, "Company Walked Out on CMA's Responsible Care Program" (Rayport 1991).

Based on its rhetoric, the chemical industry appears serious about doing the right thing. Joseph Rees, director of Virginia Tech's Center for Public Administration and Policy, writes that the chemical industry feels the public's distrust is a "major threat" and is depending on Responsible Care to build trust. A quote that emphasizes the seriousness of the industry's commitment is, "any significant failure to deliver fully on the promises of Responsible Care would be a major threat to gaining public trust and avoiding unnecessary legislation or regulation" (Rees 1997). The CMA's main goal is to "earn the public trust." The key word here is *earn*. Other words that have been used by CMA in describing the program are *ethical framework, commitment,* and *ethic* (Rees 1997).

Further evidence of the strong commitment to Responsible Care by chemical industry management was from a February 1999 speech by Bryan Sanderson, chairman of the board of the ICCA, at the International Labor Office meeting on Voluntary Initiatives Affecting Training and Education on Health, Safety and Environment in Geneva. He emphasized that the way others see Responsible Care is important. "I want it to be seen as the very essence of our industry, and not simply a window dressing exercise designed, in calculated fashion, to improve our image. It is about performance, then the communication of that performance to our stakeholders, listening to them and responding positively to their concerns" (Chynoweth 1999).

Based on the reported results and the commitment statements by chemical industry representatives, it appears to me that Responsible Care is not a public relations ploy. There is no doubt that it is implemented to improve the industry's image; however, I feel that the statements of support are real because there are measurable environmental improvements.

As discussed earlier, in order to ensure that member companies are following the commitment to Responsible Care, verification of adherence to the requirements must occur. The primary means for verification is self-assessment — each company must evaluate its progress toward implementing Responsible Care. Some country associations are moving toward third-party reviews of their programs through the MSV (Gunningham 1995). One interesting development is evaluating a joint MSV/ISO 14001 audit option for CMA members (Springer 1996). This is intriguing because ISO 14001 requires independent auditors to annually assess the progress of registered facilities against not only the ISO management systems but also Responsible Care. If the site does not keep up with its commitments, the registrar will revoke the site's ISO 14001 certification.

Not only has Responsible Care changed the way chemical companies think about the environment; it has also impacted some environmental groups' thinking as well. David Sand, project director of the CERES (Valdez) Principles — his environmental group's voluntary management program for industry — claims Responsible Care is comparable to his group's system. Although it does not include outside people, "it is an industry-specific iteration of what a company would need to do to live up to the spirit of the Valdez Principles" (Rayport 1991).

Benefits of Responsible Care

- Improved public image/community relations
- Improved environmental performance
- Thorough self-regulatory program
- Corporate executives are more supportive of environmental issues
- Publicly report progress toward initiatives
- Verification program

Some environmentalists have gone even farther in their support for this program. Roger Pryor, executive director for the environmental group Coalition for the Environment, believes that it is a "serious, even heartfelt commitment" (Rayport 1991). He served 7 years on the CMA's Responsible Care public advisory panel, a group of voluntary nonchemical industry people that advises the directors of the program (Mullin 1998). Pryor feels that reforming industry from within makes a lot more sense than the "we/they" mentality of environmentalists (Rayport 1991). Therefore, it appears that in some instances the program is getting the public to feel that the chemical industry is trying to do the right thing.

B) *Criticism of Responsible Care*

Much has been written about the benefits of Responsible Care; however, there are also criticisms. The chemical industry has critical comments about its own program. The ACC states in its 2000 Responsible Care Progress Report that "as an industry, we do a pretty poor job of communicating with the public on a range of issues." (ACC 2000). The International Association claims that its opportunities lie in improving stakeholder dialogue, performance measurement, and reporting. It states that "though we have made significant progress in dialogue with stakeholders ... more must be done." Also, "more widespread performance measurement and reporting" must be done because "our credibility demands continued progress in openness and transparency about industry performance." Further efforts are required to get small- and medium-sized companies to sign on and strive for continuous improvement and to have an open dialogue with the public about sustainable development issues, linking society, economy, and the environment together (ICCA 1999).

It is obvious that the objective of building trust in the industry has not been fully achieved. You need look no further than the results of a June 1999 survey performed by Earnscliffe Research & Communications for the Canadian Chemical Producers Association (the oldest active Responsible Care program). "The industry's reputation has improved since 1986 but still suffers from some important weaknesses." There were more *fair* and *poor* answers than *good* and *excellent* "in all but three of eleven categories probed." Most significant was that the worst-rated category was "being honest with the public, where only 18% of respondents rated the industry *excellent* or *good*" (Schmitt 1999). To further describe the lack of progress in this area, a CMA survey taken 10 years after initiating Responsible Care resulted in only 24% of the public stating, "the chemical industry protects health and safety," and only 22% agreed that firms are "accessible and willing to talk to the public" (Mullin 1998).

The lack of third-party oversight is one of Responsible Care's most serious shortcomings.

The critics of Responsible Care are most vocal about the vagueness of the program and the lack of third-party verification. There is no one uniform way to judge the achievements of companies toward the program goals because each country can set up its own indicators. For example, in the U.S., progress toward the pollution prevention code is measured by data reported to the EPA, the Toxics Release Inventory. This is a U.S. requirement that other countries such as the United Kingdom do not require. The vagueness can be easily overcome by having clear, measurable objectives. "Believable measurements of performance are absolutely crucial to the success of the program" (Ember 1992). John Ehrenfeld, director of MIT's technology, business, and environment program, believes that the principles of Responsible Care are sound but that they are also "far too vague and imprecise to lend credibility to the program and gain the trust of the general pubic" (Mullin 1998).

Shortcomings of Responsible Care
- Vague requirements with no specific goal
- No third-party verification
- Internal and external communications are lacking
- Small companies have difficulty implementing the program
- Some environmentalists view it as a public relations scam
- The concept of sustainability is not embraced

Another way to build confidence in the program, besides having clear and measurable objectives, is assurance that companies are meeting the voluntary standards. Gunningham argues that the lack of third-party oversight is "one of Responsible Care's most serious shortcomings." The entire program is based on individual companies abiding by the codes and sending performance self-assessments of how they are doing to the national associations. Therefore, monitoring and enforcement is "exclusively based on self-monitoring and self-reporting." Gunningham believes that a tough inspection and enforcement structure will "clearly enhance both the effectiveness and the credibility of Responsible Care." He goes further in stating that the administration of the program should be separated from the lobbying activities of industry associations. The ICCA and national associations should also consider some type of alignment with government to form a "co-regulatory" arrangement. He feels so strongly about having third-party verification that he claims the program is unlikely to "substantially improve the environmental performance of the chemical industry, or regain the trust of the public" without a "co-regulation scheme" (Gunningham 1995).

The need for third-party verification has been discussed by industry associations. The Canadian Chemical Producers' Association, the national chemical industry group, is evaluating a verification system with some of its member companies. This system would include independent third-party auditors to evaluate the progress of specific Responsible Care member sites in achieving the program goals. Some U.S. companies are also pursuing outside verification. The methods under evaluation include "one-day verification visits to plants by independent industry and community members, consultants' overviewing programs, the formation of community advisory panels to examine safety audit results, and the creation of independent assessment teams headed by a credible community representative" (Gunningham 1995).

There is significant disagreement among chemical companies, however. For example, in Europe, Gary Dukes, operations manager at A.H. Marks, believes that the current requirement of self-assessment is just a good start. He feels the added credibility of third-party verification is necessary because it can "bring competitive advantage" to a company providing a way to distinguish itself as "the better company." Opposed to this view are Shell HSE's manager Jo Draper and Peter Donath, head of environment health and safety at Ciba, the Swiss specialty chemical company. They even dispute the importance of self-assessments, never mind external third-party verification (Hogge 1999).

I would agree with those who believe third-party verification is an important step for Responsible Care. Having an independent group, such as an environmental consultant or an ISO 14001 registrar (an independent body approved to issue certificates to companies that conform to the requirements) will add credibility to and public confidence in the program because it is not only the chemical company auditing itself.

As for public confidence in the program, perhaps the most obvious criticism that an environmental group could make about Responsible Care is that it is nothing more than a public relations ploy by the chemical industry. Florence Robinson of the Louisiana Environmental Action Network claims that the program is "basically a PR initiative." She claims that "community advisory panels engage in mis-education or brainwashing by using the ill-defined criteria of sound science to justify industry practices and keep critics at bay" (Mullin 1998). Alluding to the same concern, the U.S. Public Interest Research Group (PIRG) laments that the U.S. chemical industry actually increased the amount of hazardous waste managed by 1.9% from 1991 to 1996 according to data reported to the government. This is contrary to the Responsible Care goal of reducing the impact of the industry on the environment (Mullin 1998).

Perhaps the most cynical view of the program was presented in the environmental advocacy magazine *Mother Jones*. It claims the CMA has "launched a fifty-million-dollar, five-year national publicity campaign called Responsible Care." Taking up a magazine ad's suggestion to "find out what your local chemical company is making," reporters contacted several chemical companies, such as Dow, Du Pont, and GE Plastics, and some comical conversations ensued. At Du Pont, the reporter was referred to "a senior public-relations representative" who did not return calls and sent an annual report without a return address or phone number. Dow did no better; when asked about toxic chemical production and waste records, it replied, "Why do you want to know?" Upon contacting GE about Responsible Care, *Mother Jones* was told, "I don't know what you're referring to." The magazine also claims that the whole program is a public relations campaign because — at the same time magazine ads were released touting the new, responsible chemical industry — the CMA was "testifying in Congress against the Community Right-to-Know More Act" (Anonymous 1992).

Further, a 1998 phone survey of 187 U.S. chemical plants in 25 different states performed by PIRG provided surprising results. Most companies were "not willing or able to share basic information with the public," and PIRG representatives could not obtain answers to "a list of seven questions about toxic chemical use and accidents." These survey results are similar to findings in a 1992 PIRG phone survey (Mullin 1998).

The chemical industry acknowledges that it has much to do to improve stakeholders' understanding of the program, including informing its employees. Trade unions have been skeptical of Responsible Care. Although the industry made attempts to make the general public aware of this program, somehow the local employees were overlooked (Chynoweth 1999).

Additional evidence of poor communication throughout the chemical industry surfaced in a 1998 MIT study:

> Senior management had not been actively involved in communicating a commitment to Responsible Care to either employees or the public. Little or no training on Responsible Care practices was given to employees.... Furthermore, Responsible Care as an initiative was not widely recognized. Very few people could give an overview of the Responsible Care Principles" (Howard 1999).

The study revealed that even a Responsible Care coordinator confessed that he probably couldn't give an overview of the program (Howard 1999).

It is no wonder that the public does not understand the improvements being sought through Responsible Care. Without clearly communicating the benefits of Responsible Care and the steps taken to improve the performance of the industry, the public still thinks of the accidents that occurred in the past.

Another frequently cited basis for criticism is the difficulty small companies have in abiding by the Responsible Care standards. Frauke Druckrey, coordinator of Responsible Care for the Association of the German Chemical Industry, mentions that one of the challenges in her country is to get small and medium companies assistance to help meet the requirements of the programs. She claims that the staff in small companies is "just about sufficient to comply with legal requirements." They are "often forced" to turn to external experts to help them abide by the law (Druckrey 1998).

Apparently some companies are overwhelmed by the program. Facilities exist that do not have enough resources to implement all the requirements, so the plant manager or the company president must try to implement the requirements. One company complained, "there are too many initiatives coming at us all at once" (Ainsworth 1993). Smaller companies feel that too many burdens have been placed upon them. There are so many requirements that they would like the implementation process to have longer time frames so that they can achieve the program's targets (Ainsworth 1993).

Smaller companies suffer with fewer resources and feel "threatened" by Responsible Care. First Chemical Corporation, a CMA member with only five plants and 400 employees, was one of the first small companies to back Responsible Care. The company president, Ed Wall, admitted that it was difficult. Since he cannot afford to maintain a corporate staff, he had to "commit a number of his company's managers ... and to make Responsible Care part of all managers' job descriptions" (Rayport 1991). Gunningham also comments that small and medium companies are not as eager as the larger, more visible companies to embrace the voluntary initiatives. Small and medium companies are "economically marginal" and "reluctant to sacrifice short-term profit ... for the sake of long-term credibility and viability of the industry." Because they are not prominent in the public's eye, an explosion in another part of the country does not affect them; they have no "public image to protect" (Gunningham 1995). Obvious resource drains exist in these small companies when they decide to implement Responsible Care.

Sustainability has been given a lot of credence as the way in which the world should evaluate anthropogenic activity. The concept is providing for the present without sacrificing the needs of future generations. In other words, let us not use up all the natural resources today and leave none for our children and grandchildren. Responsible Care has been criticized because it "does not require companies to think about sustainability." Nash and Ehrenfeld of MIT have doubts that the voluntary initiative "will eventually lead firms in the direction of sustainability ... and raise the consciousness of managers enough that they routinely consider the environmental sustainability of their actions" (Nash 1996).

Further supporting this argument is Tell Muenzing of the U.K. consulting firm and think tank SustainAbility. Critiquing the program, Muenzing states that to improve, Responsible Care must "enter the debate on sustainable chemistry." The focus should not only be on environmental impacts but also on "social and economic factors" such as explaining the benefits of a company's activities to society and balancing its economic benefit with the environmental impact of its operation (Carmichael 1999).

Some have suggested that Responsible Care companies have not improved performance above companies not in the program. A 10-year study concluded that Responsible Care companies did not reduce toxic emissions at the same rate as chemical companies that did not participate in the program (Metzenbaum 2001).

Synopsis

Now that we have taken a good look at Responsible Care, what has been learned about its positive attributes and deficiencies? It is clear that the chemical industry had to do something about its public image in light of the many terrible accidents that occurred and the negative publicity received. Responsible Care seems to be an admirable attempt to try to do the right thing. Although survey results show minimal improvements in the general public's perception of the chemical industry, it is clear that there are pockets of success in improving the industry's image, especially in the local community. The formation of national and community advisory panels has improved the communications and trust with the local community. The fact that community members are permitted to participate in a group that advises the company is a significant development and has helped to improve local relations.

Evidence shows that environmental performance of the chemical industry has improved as a result of the initiative. Drawing attention to itself and reporting on its performance has helped reduce chemical emissions and accidents. The Responsible Care program is a very detailed and far-reaching self-regulatory program, and it has focused the attention of top industry executives on environmental issues more than ever before. The industry as a whole is holding itself more accountable to the public by releasing reports on its progress, evaluating itself against the standard, and making more information available to the public. A good attempt has been made to enforce members' adherence to the codes via self-assessments and self-reporting. The initiation of third-party verification by some national organizations is a step in the right direction.

It is obvious that the program needs more clarity on its objectives and measures. The aim of the program is vague to outsiders. Allowing each country to customize aspects of Responsible Care adds to the complexity and transparency of the program to the public. It is even difficult to grasp the elements of the program when one reviews the annual ICCA Responsible Care Report.

The use of self-assessments is a good first step, but in order to earn public trust, the Responsible Care leadership is going to have to embrace third-party verification. Having local industry associations (whose primary role is lobbying) administer the program, rather than an independent group, is of concern. Communication of the program, both inside the chemical company and outside, has not been effective. Natural allies, company employees, are underutilized ambassadors of the chemical company.

Alignment with the program requirements is in the best interest of the larger, more visible firms. Small companies have a hard time meeting all of the requirements, and there is not as much at stake for them to comply with the voluntary code. Many environmental groups still look at this program as a public relations campaign. More has to be done to demonstrate the commitment to the program and show that actual improvement is occurring to change the minds of chemical industry critics. By expanding the scope of the program, the industry could possibly bring more environmentalists into its corner by embracing far-reaching concepts like sustainability and product stewardship.

What does the Responsible Care experience teach about self-regulation? First, a self-regulatory program must have clear goals and objectives, easily understood and transparent to the public. Involvement of stakeholders in the development of the program is a good idea and will help to address the concerns outsiders have. Public reporting of accomplishments is a beneficial communication tool. Verification that signatories are complying with the program goals is essential. It is also clear that self-verification is not a very good option. In order to gain public trust, at a minimum third-party verification is required, and perhaps alignment with the government agency in the form of "co-regulation" is necessary. A self-regulatory program, if it is to encompass all types of industry, must not be so resource intensive that it is difficult for small and midsize companies to participate.

Bibliography

Abelson, Philip H. (1992). Major changes in the chemical industry. *Science*, 255(5051):1489.

Ainsworth, Susan J. (1993). Responsible care program poses challenges for smaller firms. *Chem. Eng. News*, August 9, pp. 9–10.

Anonymous (1992). Why do you want to know? *Moth. Jones*, March/April, p. 18.

Carmichael, Helen (1999). Care in the community. *Eur. Chem. News*, April (Supplement), p. 10.

Chynoweth, Emma (1999). Embracing change. *Eur. Chem. News*, April 1999 (Supplement), pp. 15–16.

CMA (1999). Responsible care code performance measures. Available: http://www. cmahq.com/responsiblecare.nsf/, June 23, pp. 1–2.

CMA (1999). Responsible Care Performance Measures. Available: http://www. cmahq.com/responsiblecare.nsf, September 24, p. 1.

Druckrey, Frauke (1998). How to make business ethics operational: Responsible Care — an example of successful self-regulation? *J. Bus. Ethics*, 17(9/10):982–984.

Ember, Lois R. (1992). Chemical makers pin hopes on responsible care to improve image. *Chem. Eng. News*, October 5, p. 19.

Gunningham, Neil (1995). Environment, self-regulation, and the chemical industry: assessing responsible care. *Law Policy*, 17:59, 68–69, 72–73, 75, 91–92, 93–95.

Hogge, Roisn (1999). Responding to you. *Eur. Chem. News*, April (Suppl.), p. 6.

Howard, Jennifer, Nash, J., and Ehrenfeld, J. (1999). Standard or smokescreen? Implementation of a non-regulatory environmental code. Massachusetts Institute of Technology unpublished paper, pp. 3, 9–10, 12, 14, 19.

Hunter, David (1999). Aiming at the right targets. *Chem. Week*, July, p. 7.

ICCA (1999). *Responsible Care Status Report*, pp. 2–4, 7–8, 35–37, 65–66.

ICCA (2000). *Responsible Care Status Report*, pp. 2, 5–6, 13, 16, 21–22, 25, 30, 51.

Metzenbaum, Shelly H. (2001). *Regulating from the inside*, Resources for the Future, Coglionese, C. and Nash, T. (Eds.), Washington, D.C.

Mullin, Rick (1998). CAER: jump-starting community outreach. *Chem. Week,* 160(25):40, 62.

Mullin, Rick (1998). Critics look for greater commitment. *Chem. Week,* 160(25):39–40.

Nash, Jennifer and Ehrenfeld, J. (1996). Business adopts voluntary environmental standards. *Environment,* 38(1):44.

Rayport, Jeffrey F. and Lodge, G. C. (1991). *Responsible Care. Harvard Business School Case Study* (9-391-135), pp. 11, 15–17.

Rees, Joseph (1997). The development of communitarian regulation in the chemical industry, unpublished paper, pp. 39–40.

Schmitt, Bill (1999). Public outreach seeks the right chemistry. *Chem. Week,* July 14, pp. 41–42.

Springer, Judy and Roberts, M. (1996). All systems go for third-party verification. *Chem. Week,* 158(26):52–53.

chapter three

Project XL case study

Background

After evaluating the industry-initiated Responsible Care program, a study of a government-initiated program will help to evaluate differences in the program approaches. Project XL is the most mature government-initiated environmental self-regulatory program. It has been touted by the EPA as a bold step for giving flexibility to industry as part of its reinvention program; therefore, a study of this program should result in an interesting contrast to an industry-initiated program. The study will cover the history of the program, eligibility for participation, program elements, and an examination of the benefits and criticisms of the program as well as an evaluation of lessons learned about its use as a self-regulatory program.

History

In an attempt to respond to the mounting criticism and calls for change of the current regulatory scheme, the EPA introduced Project XL (eXcellence in Leadership) in the May 23, 1995, *Federal Register* (60 FR27282). Project XL "is a national pilot program to test new approaches for meeting environmental goals and responsibilities." By using "site-specific agreements," the EPA intended to gather information on ways to improve the manner in which the industry is regulated. According to the EPA, the goal of Project XL is "cleaner, cheaper and smarter ideas" for approaching our environmental obligations. The intent of the agency was to perform a limited number of experiments, each of which must have the potential to demonstrate procedures that can be applied to general environmental management in any one of four areas: "facilities, sectors, federal facilities, and communities" (EPA 1998).

There are several themes that the EPA specifically would like to see explored — source reduction, methods to minimize waste, facility-wide air emission limits, and environmental stewardship. However, it will accept any project that is innovative and can potentially shape future environmental regulation (EPA 1998). The greatest incentive for participating in the program is potential regulatory flexibility. In essence, an organization can propose a new way of achieving an environmental target that is less expensive, more efficient, and less burdensome as long as it has the superior environmental performance component.

Eligibility and program elements

Any company or facility regulated by the EPA (including federal, state, or municipal government agencies) can voluntarily propose a project as long as it meets certain criteria such as better environmental results, stakeholder involvement, innovation, and transferability to other regulated firms. One of the most important (and most debated) criteria for acceptance is the demonstration of *superior environmental performance*. The EPA clarified what it meant by this in a May 1998 *Federal Register* notice. First, environmental performance must be at least as stringent as current standards; therefore, emissions cannot exceed the normal criteria in existing regulation. This is determined by establishing a quantity of pollutants and wastes that would occur if an XL project were not implemented. Other factors include:

- Use of pollution prevention measures that eliminate pollutants at the source rather than with a control apparatus
- Continuous improvement of the applicant's operations that resulted in reduced environmental impact
- Documented, historical demonstration of environmental leadership through superior performance

Each XL participant must have specific goals to achieve. The program requires a Final Project Agreement (FPA) which has an enforceable component to be achieved, such as specific emission reduction targets. Enforcement can be carried out by a regulatory agency or a citizen group such as an environmental advocacy organization that could bring suit against the participant for failure to meet its commitments. There may also be voluntary agreements that can be part of the FPA. Failure to achieve either the FPA or a voluntary agreement will result in termination of the XL project. The EPA states that if the organization fails to "act in good faith" in reporting its commitments, a "strong response from the agency" will occur. Additionally, the EPA will not approve a project when an enforcement action is pending at a site (EPA 1998).

What Is Project XL?

- A national environmental pilot program was established to test new approaches.
- Its goal is cleaner, cheaper, and smarter ideas.
- Any regulated facility can volunteer to participate.
- Projects must demonstrate superior performance and have enforceable commitments.
- The main benefit of participation is regulatory flexibility.

The tools that the EPA and state agencies use to give regulatory flexibility with this program are "alternative permits and waiver mechanisms, interpretive statements and site-specific rules." The EPA states that use of these tools is perfectly legal as long as the conditions for participation are satisfied (superior performance and enforceable requirements). Participants have "protection from liability for non-compliance" with the regulatory requirements that the XL agreement has replaced. Each project is reviewed on a case-by-case basis, and the details of the agreement must go through all the steps necessary for rule making, public notice, and comment in the *Federal Register*. This makes the project very visible and accessible to the general public (EPA 1998).

An interesting component of the process is what the EPA calls "stakeholder involvement." Examples include "communities near the project, local or state governments, businesses, environmental and other public interest groups." The purpose of having stakeholders involved in the process is to get the "preference of the community," identify issues that may not have been thought of by the participants, and "lend credibility" to the process (EPA 1998).

The EPA has accepted various kinds of projects for participation in the program. Most deal with permit flexibility, facility-wide multimedia permits (air, water, waste requirements rolled into one permit), and reduction of chemical emissions. Others address compliance issues such as self-certification and the use of management systems to improve environmental performance.

It was the agency's goal to perform 50 XL experiments and see what resulted from the experimentation. During the first few years it appeared that achieving this goal was going to be very hard due to lack of participation. However, as of March 2001, there are 53 XL projects involving industry, federal facilities, and state or county government (EPA-XL 2001).

Since the 50-project goal has been exceeded, one may wonder what will happen next? It appears that the EPA has left the XL door open indefinitely. When addressing what happens next after achieving the 50-site goal the EPA says that the "need to test new tools and new solutions will continue" and "EPA is committed to providing the means for testing and implementing new concepts" (EPA-XL 2001). So it appears that there will be more experimentation available, at least in the short term.

Analysis of Project XL

A) Benefits of Project XL

Now that we have learned about the program elements, we will examine its successes. An examination of what the EPA has said about its effectiveness in a status report published in September 1998 is a good place to start. At the time the report was published, there were ten active XL projects and 20 that were in the approval process. The EPA chose to focus on three company projects in this report: Weyerhaeuser, Intel Corporation, and Berry

Corporation. The projects were designed with the participation of stake-holders, including state and local environmental agencies. Some of the aspects of these projects include improved pollution control, increased public access to environmental information, experimental use of environmental management systems, and flexible permits.

The report claims that projects have enhanced both protection of the environment and cost efficiency. The environment has benefited from pollutant reductions and increased recycling, and participants have benefited from cost savings, better relationships with stakeholders, operational flexibility, and reduced training costs. Stakeholders also received additional information about the environmental impact of the participant's facilities — environmental emission reports posted on the Internet, participant-sponsored educational activities for the community such as mentoring students, and environmental training for community groups (EPA 1998).

Projects have enhanced both protection of the environment and cost efficiency

---

A further evaluation of the effectiveness of Project XL appeared in EPA's 1998 Annual Report, *Reinventing Environmental Protection*, which documented some clear environmental improvements as a result of XL projects. For example, the Massachusetts Department of Environmental Protection streamlined its permitting and reporting system for "up to 10,000 small businesses." In this program, industry was required to develop performance standards and self-reporting on permits. Companies not in compliance were obligated to report on how they will achieve the requirements. The EPA claimed that this project resulted in photo processors cutting discharges of wastewater containing silver by 99%. Also, dry cleaners reduced air emissions of perchloroethylene by 43% (EPA 1999).

The EPA feels very strongly about stakeholder involvement in the XL process. In September 1998 it published a report specifically evaluating the Project XL stakeholder process by evaluating the "design and conduct of the initial XL projects to reach Final Project Agreements (FPAs)." There were two types of stakeholder involvement processes: consensus decision making (used by Intel and Merck) and public consultation and information sharing (used by HADCO and Weyerhaeuser) (EPA 1998). The study found that the benefits of the stakeholder process were "improved flexible and realistic environmental planning, involvement of all interested groups, including community and intergovernmental players, and opportunities for citizen involvement in future monitoring of project implementation." They also determined that the most effective method used was Weyerhaeuser's public consultation process. The EPA theorized that the reason for Weyerhaeuser's success was the presence of a "long-standing community–company relationship" (EPA 1998).

The agency is interested in determining the potential that an Environmental Management System (EMS) like ISO 14001 has on improving the environment. An EMS is a formal, systematic method of evaluating and reducing a site's environmental impact. The added benefit of ISO 14001 is that an independent third party can be used to certify conformance to the standard. Consequently, the agency was interested in teaming with Lucent Technology when it had an idea about how to use this system.

The EPA discusses Lucent's project in its *Reinventing Environmental Protection: 1998 Annual Report*. Lucent is using its EMS to simplify environmental permitting, recordkeeping, and reporting requirements. It has involved federal and local environmental agencies and stakeholder groups in determining the goals of its environmental program. A concerted effort was made to go "outside the plant fence to ... get comments on their [sic] EMS." Results and environmental improvements are made public. The goal of Lucent's project is to determine what the advantages are to having an EMS and if incentives should be given to those that employ them (EPA 1999).

By having outsiders involved, Lucent will improve its public image and build trust with its neighbors. This project will also shed some light on the much-discussed subject of how or if ISO 14001 can be used to improve environmental performance and its effectiveness as a self-regulatory tool.

The EPA gave an additional evaluation of the program effectiveness in a November 2000 report. By this time the agency had increased the participation dramatically. At the time of this report there were 48 XL initiatives with signed Final Project Agreements. There were 16 projects that had been in place for a year or more. The report focuses primarily on these mature projects. In this report the agency quantified some of the environmental benefits in a more pronounced way, as shown in the table below. There were significant pollutant reductions from 1997 to 2000 for five projects: Crompton Corporation (formerly Witco), Intel, Molex, Vandenberg Air Force Base, and Weyerhaeuser (EPA 2000).

Amount of Pollutants Reduced from XL Projects

Criteria	1997–2000
Tons of criteria air pollutants (NOx, SO_2, particulate matter, CO)	31,775
Tons of volatile organic compounds	4,029
Tons of solid waste recycled	10,855
Tons of nonhazardous chemical waste recycled	1,648
Tons of hazardous waste recycled	1,116
Tons of methanol reused	387
Millions of gallons of water reused	1,846

Source: EPA, *Project XL: Directory of Project Experiments and Results*, Wasington, D.C., Environmental Protection Agency, 2000, p. 4.

In addition to the benefits for the environment through reduction of pollutants, further participant financial gains were noted beyond what was indicated in previous reports. For example, Crompton Corporation *saved*

$1,010,000 from 1997 to 1999 as a result of waste minimization and pollution prevention efforts. Vandenberg Air Force Base will *save about $1.5 million over 6 years* for negotiating a cheaper air emission source test while reducing air emissions below required levels. Intel has *avoided millions of dollars in production delays* in the quick-to-market semiconductor industry by having a facility-wide permit that allows for changes and new construction at the site as long as site-wide air quality limits are met. As a result of revisions to the firm's air quality and wastewater permits, Weyerhaeuser predicts *avoiding $20 million in future capital spending* for air pollution equipment. They also are *saving $200,000 a year* by recovering lime muds and reusing this waste instead of buying new lime for use in its mill production.

The EPA claims environmental benefits through reduced emissions and more cost-effective regulation for sites that have entered the program, and there are dividends to the community by having additional information available and more interaction with regulated facilities. The box below spells out the community benefits as the EPA sees it.

Community Benefits

- More input into regulated site's environmental program
- Opportunity for more trust with the project sponsor
- Access to more information about the regulated site, including progress reports
- Benefit from community projects like computer donations and improved landscaping

---

One of the most important benefits the EPA received from Project XL was the ability to take a concept and bring it to other sites through rulemaking. The idea of plant-wide limits used by Intel, Merck, Weyerhaeuser, Imation, and Anderson will be made available to everyone. The EPA intends to publish a rule that allows facilities to establish emission caps on total air emissions. In my opinion, this is a big step in the right direction. Site-wide, not-to-exceed emission caps will give greater flexibility by allowing regulated facilities to make changes without having to get permit modifications or new permits as long as they stay under the cap. This makes a lot of sense not only to facilities but also to regulatory agencies. Why should they be bothered with mundane, non-value-added paperwork processing when a site is not going over an agreed facility-wide level? (EPA 2000).

A detailed review of the program by Lisa Lund, deputy associate administrator for Reinvention Programs for the EPA, was published by the Environmental Law Institute. In her article Lund takes us through the history of the program and makes the point that the EPA has learned a lot along the way. As she says, think tanks can easily say the agency should exchange regulatory flexibility for better performance, but we should not "underestimate how hard it has been for all participants to move these projects forward." There have been numerous bumps in the road because of unclear requirements, unproved stakeholder involvement, and concerns about the legal authority of the agency to take the initiative.

Project XL encourages more dialogue and involvement of the public, offers flexibility, and can be the foundation of a performance-based system of the future.

Overall she sees value in XL because it encourages more dialogue and involvement of the public, offers "flexibility with safeguards," and can be the "foundation for a performance-based system of the future." A significant point made in the analysis is that, of the 46 progress reports reviewed by EPA, *all* project sponsors had met or exceeded their commitments, and they had received substantial benefits beyond their expectations. Further, communities also feel they are benefiting from the initiatives (Lund 2000).

Evaluations of scholars' opinions about Project XL will help us further understand the benefits of this initiative. Jody Freeman of the UCLA School of Law performed a case study of Project XL, and one of the agreements evaluated was that of a privately held Florida-based citrus juice plant, Berry Corporation. Its plan was designed to demonstrate superior environmental performance in water use, water consumption and treatment, air emissions, solid waste, and wetland conservation. Berry had 25 individual environmental permits for such things as air emissions and wastewater discharges. A typical permit has specific requirements that must be achieved such as sampling effluents, reporting results, and operating within specific conditions. Further, each permit has a separate expiration date that requires the submission of a permit renewal application to the environmental agency. All 25 of Berry's permits were consolidated into a single facility-wide operating permit. The company collaboratively developed the permit with federal and state EPA officials. The concept of a single, multimedia permit is a departure from the traditional media-specific permit system (e.g., individual permits for air emissions, wastewater, etc.).

The project also included improvements to Berry's existing environmental program. The company committed to develop a formal EMS to reduce its environmental impacts and implement self-monitoring for compliance.

Additionally, the new permit includes environmental improvements beyond what regulations call for. For instance, water consumption will be reduced from 2.0 million gallons per day (mgd) to 1.8 mgd by the year 2001. Furthermore, Berry agreed to initiate the ISO 14000 environmental management programs to help the company meet its "policy goals by establishing a system of objectives, targets … and management controls." Accountability for this project is accomplished by regular self-monitoring and disclosure of its results. The EPA will visit the plant and determine the effectiveness of the system biannually (Freeman 1997).

The main success of this project was improved agency–company relationships: "A number of agency and company representatives agreed that the atmosphere was cooperative and conducive to building trust." During development of the project, agency officials were able to give helpful advice to the company and were given open access to the plant. This is in contrast to the typically adversarial relationship that exists when a regulator visits a manufacturing plant. Other benefits included improved environmental performance, clearer requirements, and implementation of the voluntary ISO 14000 management standards (Freeman 1997).

Further benefits of this program are evident in an evaluation of the Merck Pharmaceuticals agreement. Merck operates a pharmaceutical production plant in Elkton, Virginia, located about 2 kilometers from an environmentally sensitive area, Shenandoah National Park. The park is a Federal Class I area under the Clean Air Act's (CAA) Prevention of Significant Deterioration of Air Quality (PSD) program. The Merck facility is subject to various air quality requirements such as New Source Review (NSR) pre-construction review regulations, New Source Performance Standards (NSPS), and Resource Conservation Recovery Act (RCRA) emission requirements for process vents, equipment leaks and tanks, surface impoundments, and containers. This XL project was developed with various "active" stakeholders such as the Virginia Department of Environmental Quality, Department of Interior, Rockingham County, and the EPA. An individual facility-wide emissions cap was granted by promulgating a new site-specific PSD rule. This allows Merck flexibility to change its operations without obtaining approvals as required by the more complicated and time-consuming PSD and NSR requirements. The cap is analogous to having a bubble over the site. As long as Merck does not exceed specified limits, it can change its operations whenever it likes without first obtaining a permit (Wechsler 1998).

Merck's facility is in a unique location for an industrial site — close proximity to a national park. Typically the emission of volatile organic compounds (VOCs) will react with oxides of nitrogen and sulfur to form "ground-level ozone, otherwise known as smog." Because the facility is in a rural area, the "plant's VOC emissions cause little or no environmental damage." Therefore, increasing VOC levels without increases in nitrogen oxides "does not lead to increased ozone" (Hirsh 1998).

To understand this phenomenon, an explanation of the interaction of these pollutants is required. Typically, sunlight combined with VOC and oxides of nitrogen (NOx) emissions form ground-level ozone, or smog. The Clean Air Act requires PSD permits for companies that emit VOCs to prevent the formation of smog. However, this is not so with an increase in NOx emissions. An increase of NOx in rural areas does increase ozone formation. Regions subject to this effect are called "NOx-limited areas" (Hirsh 1998).

The Merck facility is located in a NOx-limited area; therefore, small increases in VOC emissions unaccompanied by increases in NOx emissions will not result in ozone formation. Since the PSD regulations do not consider this effect, the Merck facility would have to obtain a permit when it increased VOC emissions.

The XL project was designed to benefit from this regional environmental condition that is not available under the current system (Hirsh 1998).

Some of the specifics of the facility-wide permit are emission caps for NOx, SO_2, and PM_{10}, some of which are up to 25% below current emissions. The emissions of VOC and carbon monoxide (CO) may increase as long as the other pollutants remain below the facility-wide emissions cap. The facility also agreed to convert its powerhouse from coal-fired fuel to much cleaner burning natural gas. A major environmental improvement resulting from this project was a guaranteed reduction of 300 tons per year of criteria pollutants (NOx, sulfur oxides, particulate matter) achieved primarily through the switch to natural gas (Wechsler 1998).

Merck benefited by receiving the flexibility to introduce new processes, without having to obtain individual permits, and reduced paperwork requirements. The regulatory agencies will benefit from cost savings because of reduced resources required to evaluate permit applications and issue permits. They also have a better understanding of the site's operations and improved environmental performance, resulting in a reduction in pollutants beyond current requirements. Stakeholders have become more aware of the issues at the facility because Merck has made several technical presentations and answered any questions they had (Wechsler 1998).

Project XL has allowed companies "to find smarter ways to reduce their environmental impact than they would have achieved by merely complying with all of the existing air, water and waste regulations." Weyerhaeuser was able to reach an agreement with the state of Georgia and the EPA to prevent the purchase of air pollution control equipment and instead reduce wastewater effluence by 50%, reduce storm water runoff, and "improve forest management practices on 300,000 acres to protect wildlife." Flexibility was given to Intel Corporation at a facility in Arizona so it can quickly change its processes without first obtaining regulatory agency approval. In turn, Intel will keep its air pollutants below a capped level and send quarterly reports to the community (Howes 1998).

Benefits of Project XL

- Enhanced environmental protection
- Increased flexibility
- More information available to the public
- Stakeholder involvement
- Cost efficiency
- Smarter regulation

B) Criticism of Project XL

It was not difficult to find criticisms of Project XL. Every party interested in this initiative has some negative opinions of the program. The program got off to a rocky start with the much-publicized failure of a proposal by Minnesota

Mining and Manufacturing Co. (3M) to keep its Hutchinson, Minnesota facility's "multimedia emissions to a level below existing regulatory limits in return for waivers from the standard permitting procedures" (Dorf 1998). The company claimed that its plant was voluntarily performing better than what the regulations required; they wanted to be able to cut through the red tape required for permitting. The Minnesota Pollution Control Agency supported this project and wanted to encourage 3M to continue its proactive environmental stance (Larson 1998).

The proposal fell apart because the EPA could not guarantee 3M exemption from violations of existing rules that might occur as a result of the flexibility being requested by the company. There was also difficulty with what "superior" performance meant (Dorf 1998). All the components of an XL project were present; however, because the requirements for such things as superior performance were not clear, the company withdrew its application. In fairness to the EPA, clarifying advice — such as the May 1998 *Federal Register* notice that addresses issues such as superior environmental performance — has helped to make the requirements more intelligible. Nevertheless, some still find the criteria unclear and overly complex (Steinzor 1998).

One of the most-cited concerns is the lack of legal authority for Project XL. Many attorneys have commented on this dilemma in law journals. Site-specific rule making is one of the primary tools used by the EPA to offer flexibility. Most of the statutes the EPA enforces require the agency to write rules. The site-specific rule makings are actually "discretionary regulations" under the acts the EPA enforces (e.g., Clean Air Act, Clean Water Act).

One of the most-cited concerns is the lack of legal authority for Project XL.

Thomas Caballero, attorney for the U.S. Department of Justice, performed a very thorough legal review of Project XL and claims that no statute authorizes EPA to grant this type of flexibility. "A large amount of uncertainty exists" and it "may attract legal challenges to site-specific regulations." Also, flexible permits granted do not meet the statutory requirements for modifying existing permits. "Thus, XL participants will still be operating under their existing permits, many of which will conflict with the XL flexibility" (Caballero 1998). Therefore, uncertain legal requirements discourage participation in the XL program.

Similarly, Dennis Hirsch of the Notre Dame Law School claims that four legal actions can be taken against an XL project: "a citizen suit against the facility; a citizen suit against the EPA; a claim that the EPA's decision to lift a direct statutory requirement violates the separation of powers; and a claim that the EPA's decision to lift a 'gap-filling' regulatory requirement is arbitrary and capricious" (Hirsh 1998). With the prospect of litigation, it is little wonder that the EPA has difficulty getting more participants.

In addition to the legal exposure, putting together an XL project has proved very expensive. Intel's plan for its Arizona chip manufacturing facility gives operational flexibility by allowing production equipment changes without first obtaining approval from the agency, in return for commitments to conserve groundwater, reduce waste, and recycle materials (EPA 1998). Intel was surprised by how much it cost to negotiate the agreement with regulatory agencies and public interest groups. The company reported spending $1 million to implement its project, a sum much higher than expected. These extremely high transaction costs are a major complaint of XL participants (Beardsley 1997).

The Intel project development included local stakeholders. According to Sanford Lewis, director of the Good Neighbor Project (an environmental group that promotes corporate accountability), unlike the Merck project, Intel stakeholders were not as happy with the process. Just as the company was surprised and critical of the time and costs associated with working with environmental groups, the advocate groups were also displeased with their role. They did not feel that they were adequately represented nor brought up to speed on the technical aspects of Intel's proposal.

Meaningful public participation is the most controversial aspect of Project XL.

According to Lewis, the agreement was reported to allow Intel to "emit as much as 50 times more of some of the individual toxic pollutants previously permitted." This made environmental groups feel that Intel struck a sweetheart deal with the EPA. In addition, a coalition of more than 100 community, environmental, and labor organizations claimed that Intel's project "turns back the clock on hard-won laws that protect the environment" and have seen "Project XL as a fundamentally bankrupt, backward step." Part of the problem was the feeling of one citizen who attended the project meetings and who frequently was a lone dissenting vote. He felt that the ground rules were changed in the midst of the project development to reduce his say (Lewis 1997).

Highlighting the problems with the stakeholder process, Rena Steinzor states that "meaningful public participation" is the "most controversial aspect of Project XL." A very visible criticism of this process occurred when the Natural Resources Defense Council (NRDC), a national environmental group, sent the EPA and the White House a letter on July 1, 1996, threatening to withdraw support for Project XL. The NRDC claimed that the stakeholder process was neither balanced, inclusive, open, nor centered on "consensus-based decision making." Although the EPA has made serious attempts to improve this process, it still allows project sponsors to prevent stakeholders from "vetoing proposals." This makes stakeholders feel that their opinions are not taken seriously and tends to weaken support for Project XL (Steinzor 1998).

Echoing the importance of stakeholder involvement, the EPA, evaluating its own program, states that "building trust is critical." Face-to-face meetings, early inclusion of interested parties, and site visits are ways to build the trust necessary for meaningful involvement. Input from local stakeholders and national nongovernmental organizations needs to be obtained early in the process. If they are not involved early, "the negotiation process is hampered." Also, the EPA recommended that advocacy groups receive resources to "assess technical and environmental issues" (EPA 1998).

In order for industry to consider participation in a new program, the EPA has to attract them. It appears that even though the XL *cleaner, cheaper, and smarter* mantra has some appeal, the experiences of participants have not lived up to their expectations. If there were an obvious advantage to participating in Project XL, there would have been far more companies participating in the program during its first few years. Industry's reasons for participation are "saving money" on production or compliance costs via flexibility and "keeping the company's image from becoming tarnished in the public eye." Although the proposals offer the hope of gaining these advantages, the incentives to participate are not convincing due to other problems with the program, such as legal uncertainty and high transaction costs.

In concert with the lack of incentives, several participants have complained that "EPA is too enforcement oriented for a voluntary program." Some companies withdrawing their proposals have cited "conflicts with the EPA as a major issue." Likewise, the Environmental Council of the States (ECOS) passed a resolution stating that EPA mishandled Project XL. It appears one of the council's main complaints is EPA's lack of concern for differences between a state's approach to environmental protection and EPA's concerns with state-sponsored audit privilege policies (policies giving industry enforcement variances for self-disclosed violations) (Morelli 1999).

Criticisms of Project XL

- Unclear requirements
- Legal authority is not clear
- High transaction costs
- Appearance of sweetheart deals with industry
- Stakeholders process is weak
- Not enough incentives to participate

Synopsis

Now that we have explored the benefits and criticisms of Project XL, what conclusions can be drawn about its effectiveness? First, I think the EPA should be commended for its bold attempt at improving the current regulatory system. It appears that the program has some merit in tapping the momentum of the greening of industry to gain environmental protection that would not occur under traditional prescriptive command-and-control regulation.

Environmental improvements have resulted from the program in the form of thousands of tons of pollutants removed from the environment. Companies are able to propose ways to address the environmental concerns at their sites with a commonsense approach and without being restricted by the current regulation. Significant flexibility has been gained by some companies that have streamlined the permitting process, allowing them to make faster changes to their processes and giving them a competitive advantage. The streamlined permitting and reduced permit requirements also resulted in cost savings for environmental governmental agencies since they have to evaluate and approve fewer documents.

The stakeholder process has made more information available to the public and has given interest groups access to industrial sites that they do not get under the current system. When it is working right, an effective project that is good for the environment, the company, and the local stakeholders is the result.

Project XL has its fair share of growing pains. It was off to a rocky start due to vague requirements. Although attempts at clarity have been made, there is still room to improve the directions for developing new projects. There is substantial fear that the flexibility given has no legal basis, and this, along with the possibility of citizen suits, apparently has scared off potential participants.

The involvement of stakeholders, although admirable, needs more thought. Trust must be developed so it does not appear that the EPA is giving industry a sweetheart deal. The lack of resources for local citizen groups to understand some of the complex issues, such as trading one toxic pollutant for another, is part of the problem.

Stakeholder involvement has to be balanced with the enormous amount of time and costs necessary to strike a deal with stakeholders and regulators. The EPA must also try to make participation in the program more attractive by increasing or enhancing incentives for industry.

What conclusions can we draw about the use of Project XL as a self-regulatory policy instrument? Project XL has taught us that, when designing a self-regulatory program, the elements must be very clear and well thought out. All legal concerns must be worked out so that no barrier to the effectiveness of the program exists. Stakeholders should be involved up-front in the design of the program to prevent the appearance of behind-the-scenes deal making. Ample incentives for industry involvement must exist in order for the program to be effective, and the EPA must get out of its enforcement mentality when teaming with industry to develop a self-regulatory program. Cost advantages for governmental agencies must be present to have a sustainable program.

Bibliography

Beardsley, D., Davies, Terry, and Hersh, Robert (1997). Improving environmental management: what works, what doesn't. *Environment*, 39:28.

Caballero, T. E. (1998). Project XL: making it legal, making it work. *Stanford Environ. Law J.*, 17(2):420–421.

Dorf, M. C. and Sabel, C. F. (1998). A constitution of democratic experimentalism. *Columbia Law Rev.*, 98(267):384–385.

EPA (1998). *Evaluation of Project XL Stakeholder Process.* Washington, D.C., U.S. Environmental Protection Agency.

EPA (1998). *Project XL Preliminary Status Report.* EPA-100-R-98-008. Washington, D.C. U.S. Environmental Protection Agency.

EPA (1998). Regulatory reinvention (XL) pilot projects. Available: http://www.yosemite.epa.gov/xl/xl_home.nsf/all/frn-4-23-97.html December 17.

EPA (1999). *Reinventing Environmental Protection: 1998 Annual Report.* Washington, D.C., U.S. Environmental Protection Agency.

EPA (1999). XL at a glance. Available: http://yosemite.epa.gov/xl/xl _home.nsf/all/xl_glance, August 31.

EPA (2000). *Project XL: Directory of Project Experiments and Results.* Washington, D.C., Environmental Protection Agency.

EPA-XL (2001). Project XL implementation and evaluation. Available: http://www.epa.gov/projectxl/implemen.htm. March 29, p. 7.

EPA-XL (2001). Project XL frequently asked questions. Available: http://www.epa.gov/projectxl/faqs.htm. March 29.

Freeman, J. (1997). *Collaborative Governance in the Administrative State.* Los Angeles, UCLA, pp. 57–61.

Hirsh, D. D. (1998). Bill and Al's XL-ent adventure: an analysis of the EPA's legal authority to implement the Clinton administration's project XL. *Univ. Ill. Law Rev.*, 1:133, 144, 145.

Howes, J., Dewitt, J., and Minard, Jr., R. A. (1998). Resolving the paradox of environmental protection. *Issues Sci. Technol.*, Summer, pp. 59–60.

Larson, P. (1998). A culture of innovation. *Environ. Forum*, September/October, p. 26.

Lewis, S. (1997). Feel-good notions, corporate power and the reinvention of environmental law. *The Good Neighbor Project for Sustainable Industries*, Working Paper (March 12, 1997 Version), pp. 8–10.

Lund, L. C. (2000). Project XL: good for the environment, good for business, good for communities. *Environ. Law Inst.*, 30 ELR 10140 (2-2000), pp. 3–5, 9, 11, 13.

Morelli, J. (1999). *Voluntary Environmental Management.* Boca Raton, FL, Lewis Publishers, p. 71.

Steinzor, R. I. (1998). Reinventing environmental regulation: the dangerous journey from command to self-control. *Harv. Environ. Law Rev.*, 22(1103):134, 142, 143.

Wechsler, B. S. (1998). Rethinking reinvention: a case study of Project XL. *Environ. Lawyer*, 5(1):270–275.

chapter four

OSHA Voluntary Protection Program (VPP) case study

Background

After evaluating environmental industry- and government-initiated programs, we will now study a successful government-run self-regulatory program from a related field, the OSHA VPP. This program also has over a decade more experience than Project XL. The VPP has some very attractive components that could be transferred into the environmental field, and the impetus behind its development also parallels the problem EPA and state environmental agencies now have — a need to move in a more collaborative direction to obtain better results. This case study investigates the history of the program, program elements, eligibility for participation, and an examination of its benefits and criticisms. Conclusions are drawn on what can be learned from the VPP principles to help us determine which of its aspects can be used to devise an environmental self-regulatory program.

History

The Occupational Safety and Health (OSH) Act of 1970 was enacted to "assure safe and healthful working conditions for the workers of America" (Mintz 1984). The Occupational Safety & Health Administration (OSHA) was formed to ensure that the goal of the Act was achieved. Nine years after the development of the Act, a Library of Congress report on workplace safety indicated that there were some disturbing statistics. From 1972 to 1979, lost workdays from serious injuries increased 34%. The then-assistant secretary of labor, Throne G. Author, claimed that neither management nor labor was satisfied with OSHA's implementation of the law. He felt it was time for a different direction to be taken to improve workplace safety — cooperation of government with management and labor.

Author viewed OSHA as a policeman having an adversarial relationship with the private sector. He intended to use all of the authority given to OSHA under the OSH Act: "education, training, consultation, employer–employee voluntary cooperation, self-inspections, standards-setting, and enforcement" (Mintz 1984). In the past it had only focused on enforcement and standards-setting and not on partnering with those companies it regulated. The new vision of cooperation to improve the safety of the American worker resulted in the development of the self-regulatory program called the Voluntary Protection Program (VPP) in 1982 (Mintz 1984).

OSHA has had over 19 years of experience with the VPP — more than any environmental agency-run self-regulatory program in the U.S. The EPA has come to the realization that it must try new methods to improve the environment, and it is in a similar position as OSHA was when it developed the VPP. Therefore, an evaluation of the effectiveness of this voluntary program, with its significant history in an area related to environmental protection, should result in relevant information about the potential of industry self-regulation. In order to evaluate this initiative we first address the aim of the program.

OSHA initiated the VPP because it believed that compliance enforcement alone can never fully achieve the objectives of the Occupational Safety and Health Act.

The VPP allows "a select group of facilities that have designed and implemented outstanding health and safety programs" to partner with the regulatory agency to "go beyond OSHA standards" to achieve greater protection of their workers. OSHA says it initiated the program because "compliance enforcement alone can never fully achieve the objectives of the Occupational Safety and Health Act." (OSHA 1999). Therefore, the agency believes that a cooperative approach is a more effective way to protect the safety and health of the nation's workers.

Further insight into the development of the program is given by Charlotte Garner, a safety specialist at NASA's Johnson Space Center, and Pat Horn of the OSHA Policy Office, who worked on the development of the VPP. In their very comprehensive guidebook for implementing the VPP, they state that "after ten years of adversarial compliance and many expensive court battles," OSHA realized that there were still many unsafe workplaces. OSHA was making some headway with its current policies, but worker deaths, injuries, and illnesses were occurring at unacceptable rates. Though penalties for violations "increased to astronomical figures," the fines went into the national treasury where they did nothing to protect worker health. Through a pilot program in cooperation with the construction industry in California, OSHA discovered that "companies with exemplary safety and health programs, regardless of regulatory requirements, also had low incident rates."

Armed with this new insight, OSHA developed a new cooperative program called the VPP (Garner 1999).

OSHA claims that its legal authority to initiate a voluntary program stems from the goal of the 1972 Safety and Health Act, which is "to assure so far as possible every working man and woman in the Nation safe and healthful working conditions." Section (2) (b) (1) of the Act directs OSHA to "encourage employers and employees in their efforts to reduce hazards, institute new programs, and perfect existing programs for providing safe and healthful working conditions." The goal of the VPP is to encourage companies to go beyond the regulations in cooperation with the agency (OSHA 1996).

> The goal of the VPP is to encourage companies to go beyond the regulations in cooperation with the agency.

Program elements

Companies must go *beyond* the requirements of the safety and health regulations to participate in the VPP. The program requires the "implementation of management systems that provide integrated, comprehensive, and ongoing safety and health protection to workers" that exceeds the requirements of OSHA (Catanzaro 1994). There are 19 elements that OSHA requires participants to meet to become a VPP site. Each one of the following elements has procedures and requirements to which a participant must adhere.

1. Management commitment and planning
2. Accountability
3. Disciplinary program
4. Injury rate
5. Employee participation
6. Self-inspections
7. Employee hazard reporting system
8. Accident/incident investigation
9. Job safety analysis/process reviews
10. Safety and health training
11. Preventive maintenance
12. Emergency programs/drills
13. Health programs
14. Personal protective equipment
15. Safety and health staff involved with changes
16. Contractor safety
17. Medical program
18. Resources
19. Annual evaluation

Some of these elements are explicitly required by OSHA regulations, such as the necessity for employers to provide safety and health training and personal protective equipment to workers. Others, such as holding line managers and supervisors accountable for safety and health and having a written disciplinary program for violating company safety requirements, are not regulatory requirements (Garner 1999).

Details of one of these requirements is illustrated by some of the questions that OSHA VPP employees are asked during an on-site evaluation of a participant program.

VPP on-site evaluation questions for management leadership

1. What management commitment to safety and health protection did you observe?
2. What evidence did you see that established policies and results-oriented objectives for worker safety are still being communicated to all employees?
3. What evidence did you see of an established goal for the safety and health program and objectives for meeting the goal?
4. Are the goals and objectives communicated effectively so that all members of the organization understand the results desired and the measures planned for achieving them?
5. Are authority and responsibility for safety and health integrated with the management system of the organization?
6. Has management shown a clear commitment to maintaining the requirements of the VPP? How? (OSHA 1996).

As evidenced by these questions, many beyond-compliance elements are contained in the VPP.

Eligibility

The program has three levels — the highest is called *Star*, then *Merit*, and finally *Demonstration Program*. A Star company must have "occupational safety and health programs that are comprehensive and successful in reducing workplace hazards." Therefore, OSHA requires superior programs to be in place. Examples of elements that indicate superior performance include having:

- A safety and health evaluation system in place for at least a year
- A documented system of management accountability from the CEO to the first-line supervisor in place for at least a year
- A 3-year average lost workday injury case rate and a 3-year injury incidence rate at or below the national average for the Standard Industrial Classification (SIC) code for the industry.

The key for determining if a company should be in the program is whether it is operating at Star quality (OSHA 1996).

A Merit company is one with good programs that are not quite up to the Star level. OSHA and the company create an action plan of "stepping-stones" to become a Star performer. A Demonstration Program allows evaluation of criteria different from, but potentially as protective for workers, as the Star criteria. This is a program to "demonstrate" that these criteria also protect workers (OSHA 1996).

The process for becoming a VPP company is arduous, and OSHA wants it to be that way.

The process for becoming a VPP company is arduous, and OSHA wants it to be that way. "The VPP application process is designed to be rigorous, to assure that only the best programs qualify" (OSHA 2001). The application is very detailed and requires such things as injury statistics, commitment from management to "provide outstanding safety and health protection" to employees, and a statement from the labor union supporting the VPP application. A description of the employees' involvement with safety and health programs, documentation of the existence of job hazard analysis, self-inspection, and accident investigation programs are also requirements (OSHA 1996).

OSHA initially verifies that the program meets the VPP criteria with a preapproval on-site review. If the firm is accepted into the VPP program, it is publicly recognized via a press release that lists all the VPP companies. The site is then removed from "routine scheduled inspection lists," although if a major accident, employee complaint, or chemical spill occurs, a traditional inspection will take place (OSHA 1999).

VPP companies are removed from the routine schedule inspection lists.

To maintain VPP status a company must face OSHA verification visits every 3 years for the Star program and every year for the Merit program to confirm that these sites continue to meet VPP criteria (OSHA 1999). The VPP staff, rather than ordinary OSHA field inspectors, performs the reassessment visits. The evaluation is detailed and tedious, resulting in a formal report that indicates a site's strengths and weaknesses. The assessment team reviews various records, such as the OSHA 200 Log — which is required to be maintained by employers to track injuries — and safety and health programs (inspections, accident investigations, training, and personnel protective equipment). The team also interviews employees and tours the site. If there are any deficiencies cited, the company has to correct them within specified time frames. If OSHA and the company cannot agree on both problems and solutions, VPP status may be revoked (OSHA 1996).

During the site assessment the VPP personnel look for evidence that the company's program is "creating and maintaining safe and healthful working conditions." If a violation of an OSHA standard is recognized, the site is asked to correct it. The assessors do not issue citations; however, they will not ignore hazards. "If corrections require more time than the on-site review allows, you will be asked to notify your OSHA VPP Program Manager when corrections are completed." If a "cooperative resolution" cannot be reached, enforcement action is taken (OSHA 1999).

In addition to the federal OSHA, states that have authority to run safety and health programs can also initiate VPP programs after receiving OSHA approval. VPP membership has significantly increased since the program's inception. In 1982 there were only 11 companies in the program — in 2000 there were 543. The state-run programs started in 1992, with just two companies participating. There were 162 companies participating in 2000 (OSHA 2001).

Analysis of OSHA VPP

A) Benefits of OSHA VPP

We have evaluated the history of the VPP and discussed its purpose and elements. Now we examine what has been said about the effectiveness of the program, starting with OSHA itself. OSHA considers VPP companies as "models for their industries." Therefore, a certain amount of esteem comes

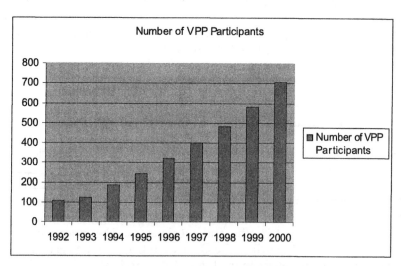

Figure 4.1   Growth of VPP. Number of VPP participants in both federal and state programs as of December 31, 2000. (From OSHA 2001, Growth of VPP, available at http://www.osha.gov/oshprogs/vpp/spgrowth.html and http://www.osha.gov/oshprogs/vpp/fedgrowth.html, March 29, 2001.)

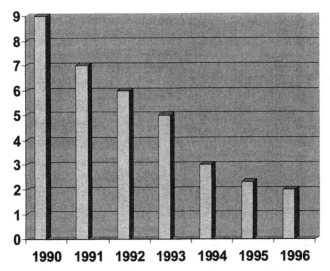

Figure 4.2 Georgia Pacific serious injuries per 100 employees. (From Fisher, A., Danger zone, *Fortune*, September, 167, 1997. With permission.)

with participation in the program. Because a company commits to cooperating with the agency, it is not subject to routine inspections. On-site reviews by VPP personnel ensure that "safety and health programs provide superior protection." OSHA claims significant cost savings because VPP companies have lower worker injury rates. *VPP companies are 55% below the typical injury rate compared to other companies in their industry group.* Additional benefits include decreased workmen's compensation costs and lost work time as well as increased employee morale and productivity (OSHA 2001).

Before we evaluate what others outside of the agency have said about the VPP, it should be noted that there were few scholarly works available in the literature. Most of the information used to evaluate the initiative is from advocates of the VPP because of their participation or administration of the program.

Significant cost savings were documented in a guide written by Margaret R. Richardson, who managed the VPP program for 8 years for OSHA. For instance, Georgia Pacific, a forest products company, had averaged nine serious injuries per 100 workers each year and 26 total fatalities from 1986 to 1990. The company embraced the VPP and had ten sites in the program by 1996. Its performance improved drastically. In 1996 it recorded injury rates *59% below* the industry average and lost workday cases (a measurement that indicates the number of days injured workers are off the job) that were *62% below* industry average. The company only had two serious injuries per 100 workers in 1996.

June Brothers, senior manager for Human Resources, claimed that *Georgia Pacific had saved $42 million in workers compensation costs* from the years 1990 to 1996, while it had been working toward VPP Star status.

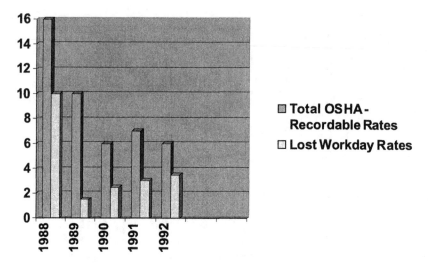

Figure 4.3   Injury rates at Southern Tool, Anniston, Alabama. (From Richardson, Margaret R. *Preparing for the Voluntary Protection Programs*, New York, John Wiley & Sons, 1999, p. 16. With permission.)

Similarly, Mobil Oil's Joliet Refinery in Illinois reduced injuries by 92%, and worker compensation costs dropped by 85% when it started working toward the VPP. Prior to its participation in the program, worker compensation costs were averaging $250,000 per year. During its first years in the VPP, 1990 to 1993, the average costs *dropped to $48,000 per year*. As the Joliet Refinery continued in the program, even more cost efficiency was gained with costs of *$20,000, $9500* and *$4000* in the years 1994 to 1996, respectively. Mobil's refinery in Paulsboro, New Jersey, had comparable success. Because of its participation in the VPP, worker compensation costs were *reduced from $200,000 in 1991 to $20,000 in 1994*.

Milliken and Company, a textile mill with 181 employees, achieved a lost workday rate 98% below industry average which resulted in savings of $238,500 per year.

Smaller firms have had success as well. Over the course of 5 years, Southern Tool, a small foundry in Anniston, Alabama, reduced recordable injuries by 60% and lost workday rates by 58% (Richardson 1999). Milliken and Company, a textile mill with 181 employees, achieved a lost workday rate 98% below industry average, which resulted in savings of $238,500 per year. ABB Air Preheater, a heat exchanger manufacturer with 611 employees, has recorded lost workday rates 73% below industry average for 15 years. This is estimated as a *$1,069,200 savings per year* (Siddiqi 1999). Therefore, OSHA and others believe that participation in the VPP yields considerable cost savings.

Next to OSHA, the biggest supporter of the program is the Voluntary Protection Program Participants Association (VPPPA). The VPPPA is an association of more than 600 companies and work sites that participate in the VPP program. Its purpose is to foster cooperative efforts among labor, management, and government in the pursuit of safety, health, and environmental excellence. The association is committed to broadcast the benefits of the VPP and to encourage and support companies in their efforts. It holds annual conferences, workshops, and meetings to assist participants in their VPP journey (VPPPA 2001).

In its publications it reiterates several themes to encourage VPP participation, the most prominent of which is the cooperative nature of the program. During the inaugural partnership conference sponsored by OSHA, titled "Partner with OSHA: New Ways of Working," VPPPA member Huntsman Petrochemical Corporation of Port Arthur, Texas, described some of the partnerships in action. Huntsman has been a VPP facility since 1987. In addition to cultivating a good relationship with OSHA, the company's program improved labor and management relations. Hermie Bundick, plant manager at the Port Arthur facility, stated that trusting each other with safety leads to trust in other areas "which leads to productivity and a safe working environment" (Phillips 1999).

Also lauding the benefits of cooperation is VPPPA member Dr. Stanley D. Weinrich, manager, Environment, Safety and Health for International Paper in Memphis, Tennessee. He claims,

> The program has broken down communication barriers that formerly existed between industry and OSHA regarding safety issues. No longer do we speak to OSHA only through attorneys and with layers of business management filtering the communication. ... hourly employees have been known to call OSHA directly to seek guidance on the best way to resolve a problem (Philips 1999).

It is apparent from the VPPPA publications that its members are clear on the benefits of partnering with the agency.

Apparently regulators and participants feel the VPP has resulted in a new paradigm — that of trust and cooperation. Jeff Johnston, Texas Eastman Division of the Eastman Chemical Company, also agrees that the "VPP removes the adversarial relationship." He feels comfortable asking OSHA contacts for advice about potential problems at his site without expecting to see inspectors reward his inquiry with a notice of violation. Similarly, Dale Saunders of Fort Howard Corporation's Savannah River Mill in Georgia says that the VPP team is there to help, making suggestions on how to improve, and that the team's visit was "educational" (Richardson 1999).

Not only does industry share in this view but agency officials also feel the same. Richardson, writing about her early experiences with the VPP, tells the story of a visit to a plant that had a "Welcome OSHA" sign at its entrance (Richardson 1999). Some members of the evaluation team were taken aback by this gesture of acceptance, which was so different from the usual reception. The agency has apparently become accustomed to warm welcomes and the shift away from being viewed as "intrusive, unreasonable and ... unhelpful" (Richardson 1999). It also appreciates seeing employers concerned for the well-being of their employees.

From an industry perspective, having a government agency come to the company's site to check out the effectiveness of its safety program could scare some away from VPP. The question that first comes to mind is, what if a violation is found? OSHA states that any hazards observed "will serve as indicators that some aspect of your program may need improvement." In the spirit of cooperation, OSHA claims that citations will not be issued for hazards, but any found will be brought to the company's attention.

> They will work with you to determine how and when to correct any hazards they see. If corrections require more time than the on-site review allows, you will be asked to notify your OSHA VPP Program Manager when corrections are completed. Should all attempts at cooperative resolution fail, the VPP team has a responsibility to recommend to the Assistant Secretary that enforcement action be taken (OSHA 2001).

This approach gives the participating industry assurance that it will not receive fines or violations for participating in a proactive program — a very important point for many companies.

Amazingly, OSHA officials have come to trust VPP sites so much that they suggest companies struggling with safety and health programs contact these sites for help. Even further trust is displayed by accepting VPP site personnel as "special government employees" (SGE) to augment VPP on-site teams and help keep VPP application backlogs down" (Richardson 1999).

The SGE program also allows industry members to visit another VPP company's site as part of the OSHA evaluation team. (The site being evaluated has to agree to the participation of the SGE prior to the assessment.) Each SGE "must complete a 3-day training course in how to assess a work site's safety and health management systems" and determine if the site "meets VPP requirements" (Sherrill 1998). The industry volunteer will then be sworn in as a special government employee. All expenses for the volunteers are paid by their employers.

Apparently this aspect of the program is invaluable to the agency. Susan Sikes, OSHA's VPP manager in Region IV, Atlanta, states that "we could not continue to run a successful program without our VPP volunteers." This effort is helping OSHA expand the VPP while conserving its limited resources. "In 1993 more than 50 VPP Volunteers from 40 companies participated in more than 100 on-site reviews" (Sherrill 1998).

Additionally, VPP site employees have helped train OSHA employees. Extensive training was provided by industry employees during a 6-week Advanced Level-1 Process Safety Management (PSM) course for OSHA compliance officers (Catanzaro 1994). I am not aware of any other program that allows industry to participate in program oversight and to instruct agency employees. This type of cooperation between government and industry is atypical and evidently spreads good will to all parties. To Richardson, a further benefit of the cooperative nature of this program is the improved relationship between employees and management. This is best highlighted by looking at how the VPP has affected the typically adversarial union/management relationship. Union work sites "made up 27% of all work sites" in the VPP in January 1999. Randy Little, chief steward for the United Food and Commercial Workers Union Local 1450, speaking about the poor relationship at Armour-Swift-Eckrich in St. Charles, Illinois, said, "it used to be that nobody could stand each other, that nobody wanted to communicate. VPP has helped give us a common goal." Shirley Lockwood of Lucent Technologies in Shreveport, Louisiana, and of IBEW Local 2188, said that her work site had a "we/them" way of thinking. The company and the union decided to get involved with VPP anyway. The training and team building that resulted helped them to work together and "gave everyone a voice" (Richardson 1999). VPP has improved union/management relations at several companies.

Once a site receives VPP Star status, it is immediately recognized as having achieved a superior level of performance. This recognition gives it immediate equity with the community and its peers because it is a widely recognized standard of excellence. The VPP site is expected to participate in outreach activities, such as those that encourage the improvement of safety and health, and the mentoring of other industrial sites (Weinberg 1997). Paul Varello, chairman of American Ref-Fuel, a waste-to-energy company, welcomed participation in the VPP because it was an "indicator of our company's commitment to safety." It is a certification that "bolsters public confidence" and accentuates the company's commitment to being a "good corporate citizen" (Siddiqi 1999).

> The VPP is a certification that bolsters public confidence and accentuates the company's commitment to good corporate citizenship.

Once a company is accepted into the VPP, an approval ceremony occurs at the facility. In addition to company representatives, local, state, and national officials usually attend. The company is presented with a certificate of accomplishment and a VPP flag, symbols of outstanding performance. Some companies even take "full-page ads in their local newspapers" to broadcast their achievement (Weinberg 1997).

Neil Gunningham and Richard Johnstone of the Center for Employment and Labor Relations Law at the University of Melbourne have written a scholarly evaluation of workplace safety law. They comment positively about the VPP. Besides the benefits of improved safety and reduced worker compensation costs, there is a public relations value in receiving recognition for outstanding programs and use of the Star logo (Gunningham 1999). The commitment to go beyond regulatory requirements and partner with agencies increases community confidence in a company's operations.

Most companies set their safety and health goals to meet OSHA regulatory requirements. The VPP program not only requires conformance to regulations but goes much further. An example of this is the systematic way a company must implement safety programs. The VPP method necessitates "management leadership, employee involvement, hazard assessment, hazard prevention and control and employee training." This is comparable to a quality management system for a company's safety and health program. Management leadership and employee involvement are considered key elements for world-class safety programs.

Management leadership is one of the keys to achieving world-class safety (Minter 1997). The success of any business initiative requires significant management support to be prosperous. Many safety professionals want to see more management involvement in their programs, and the VPP is a perfect venue for doing this. Requirements such as having goals and objectives, assigning responsibility, and providing resources are examples of how management must be involved. Since management controls resources and time, having its support really makes worker safety happen.

Benefits of VPP

- Provides better safety and health for employees
- Results in cost savings for participants
- Develops good relationships with regulators and employees
- Enhances public image for the company
- Forces a systematic safety and health program
- Decreases inspections and penalty mitigation

Employee involvement also has clear advantages. The total quality management movement demonstrated the gains from empowering employees. The VPP requires the support of all workers. In fact, without union support, the program cannot be implemented. Once a company has its workforce behind the program, it is so much easier to achieve safety excellence. There are more "eyes" looking for hazards and correcting problems. The safety system puts the company in a prevention-based approach rather than a reactive one (Richardson 1999).

OSHA actually helps establish the safety and health system at a company's plant when it performs verification visits at a participant's site. The VPP guidelines are a proven how-to guide for effective programs. These guidelines reinforce the importance of the safety program and give feedback on its effectiveness (OSHA 1999). A systematic program that demands routine checks from a friendly outside agency ensures that good programs will continue and gives management assurance that safety and health issues are handled properly.

John C. Ruhnka, professor of management at the Graduate School of Business Administration at the University of Colorado at Denver, considers the VPP a leading example of regulatory certification incentives because it gives "promised relief from certain routine OSHA enforcement requirements, a promised closer working relationship with OSHA, and expedited response to requests from the agency" (Ruhnka 1998). The agency claims not to subject participants to "routine OSHA inspections, because OSHA's VPP on-site reviews ensure that their safety and health programs provide superior protection" (OSHA 1999). The VPP reviews are not enforcement oriented like a typical compliance inspection because the assessors are working as partners in the spirit of the cooperative program. Participants receive significant regulatory incentives because they do not have routine inspections and are not issued citations for violations noticed during verification visits.

OSHA believes it can use its experience with the VPP to encourage poorly performing industries to improve. By using statistics such as the OSHA 200 log and workmen's compensation claims, the agency is able to identify those industries with the greatest number of injuries. Companies that have been singled out are given a choice of either being subject to "the highest priority of traditional OSHA enforcement" or partnering with the agency to implement a "comprehensive safety and health program." OSHA calls this the Cooperative Compliance Programs (CCP). Most companies choose the cooperative approach, which has provided impressive results in injury reduction and worker compensation costs (Weinberg 1997).

Therefore, the voluntary approach can be used in a positive way to improve companies with poor records, as compared to the traditional command-and-control enforcement approach.

B) Criticism of OSHA VPP

One of the most cited problems with participation in the VPP is the complexity and increased burden upon entering the program. We need not look any further than the seven chapters and 218 pages of the *Revised Voluntary Protection Programs Policies and Procedures Manual*. When a company decides to participate in the VPP, it has to conform to OSHA regulations and must also take on the extra burdens of the VPP.

The program is so complex that it usually takes a company from 2 to 5 years to fully implement all the elements. Some companies have even taken as long as 7 years to conform to the system. Depending on the maturity of an existing safety and health program, there could be a significant cost to participate as well. There are increased training, analyses, reporting, inspection, and paperwork demands that go above and beyond regulatory compliance requirements. Therefore, a company may have to increase its safety budget to be a participant (Garner 1999).

The application process alone can be arduous. OSHA wants a complete picture of the site's safety program. There is a considerable amount of detail required in the eight sections of the application. Besides general company information and injury and illness rates, detailed descriptions of such information as the company's safety policy, organizational structure, safety responsibilities, resources, and integration of safety must be included. Details on health and hygiene surveys, self-inspections, job hazard analyses, accident investigations, on-site medical care programs, safety rules, and training programs are also required. Even after putting all this information together, an application can be turned down or additional information requested (Swartz 1995).

Nor is it getting any easier for companies to participate. A 2000 *Federal Register* Notice revising the program proclaims that the new criteria will "make the VPP more challenging and ... raise the level of safety and health achievement expected of participants." There are new illness reporting criteria and more requirements for contractors working at VPP sites (OSHA 2000).

The fact that the VPPPA holds conferences to familiarize and train companies interested in the program indicates that this is not a simple process. The celebration associated with achieving Star class is for a good reason. It is not easy to fulfill all of the requirements of the program. Star applicants must have OSHA recordable injuries and lost workday cases that are at or below industry average. This is a beyond-compliance program, and it is not for companies that do not have good safety programs already in place.

VPP program supporters make it clear that this program is not for marginal performers; yet they quote the great results that companies achieve from being in the program — injury rates 55% below industry average along with tremendous cost savings (OSHA 2001). Gunningham has suggested that it is unclear if these stellar results are from the VPP or because the companies allowed to participate already had good safety programs (Gunningham 1999). This argument has merit because of the very strict entry requirements. The agency says that it is looking for work sites that have strong programs (OSHA 1999). This prompts the question: if this program protects employees so much, why doesn't OSHA let poorly performing companies in too? Don't these companies need the extra help more than the outstanding companies do?

Some people cringe at the idea of inviting a government regulatory agency to their site. After an application is received, up to three compliance officers visit the site for several days to verify that programs are as indicated in the application and that potential hazards at the site are addressed. Though OSHA states that violations will not result in fines or citations if promptly corrected, a company does bring increased scrutiny to its site (Swartz 1995).

When I first learned of this program, I was most surprised that in addition to submitting annual data to OSHA on safety and health programs, a company subjects itself to routine periodic reviews. It is common knowledge that OSHA rarely visits work sites unless it receives a complaint. It has far more sites to inspect than the EPA does and a smaller staff to do the inspections. "In most industries, OSHA inspects less than 10% of all work sites annually." Likewise, "decades can go by before an OSHA inspector knocks on your door" (Siddiqi 1999).

Richardson even states that "VPP sites are visited by representatives of the regulatory agency more frequently than if they were not in the program." If a company is in an industry that does not have high incident rates, there is little reason to join the VPP to avoid normal OSHA inspections because the agency schedules visits based on an industry's safety record (Richardson 1999). The level of scrutiny to which a VPP participant is subjected poses another question: Is this the best use of the agency's personnel? A Star company is evaluated every 30 to 42 months with a maximum 60 months between visits. A Merit company is visited every 24 months (OSHA 1996).

> Why should the agency spend significant amounts of time on sites that are committed to superior performance and overlook the hundreds of thousands that are not?

Is OSHA using its resources wisely? Why should the agency spend significant amounts of time on sites that are committed to superior performance and overlook the hundreds of thousands that are not? OSHA addresses this issue by stating that VPP visits are more efficient than routine inspections. In 1992, "2.46 full-time equivalent OSHA safety compliance specialists and 2.46 full-time equivalent OSHA industrial hygienists were used to conduct the 65 pre-approval and participant evaluation VPP sites. Using OSHA data, 196 safety inspections and 65 health inspections could have been conducted if these employees had not been on VPP on-site visits" (OSHA 1999). Nevertheless, the efficient use of resources and even the innovative SGE program does not diffuse the argument that the OSHA staff should be concentrating on the bad actors — not the stellar performers.

Perhaps another way to ensure that superior programs continue at the VPP sites after they are accepted into the program is third-party certification. Rather than use agency personnel to perform site visits, independent consultants, paid for by the participants, can be used to verify continued conformance to the VPP standards. Reports by the consultants could be sent to the agency for review. This should not add that much to the cost of the program for participants (I would estimate it would cost under $8,000 every 3 years); further, it will help to focus OSHA resources on the poor performers, which should result in a higher level of protection of more workers.

In order to participate in the program, the site must have the support of any collective bargaining unit present. Therefore, if the local labor union does not want to participate, it can prevent the company from gaining the benefits of the VPP. Part of the application requires a statement about the union's stance.

Union Statement. If the site is unionized, the authorized
collective bargaining agent(s) must sign a statement to
the effect that they either support the VPP application
or that they have no objection to the site participating
in VPP. The statement must be on file before the application is considered complete (OSHA 1996).

This can be a stumbling block because misunderstandings can prevent
a well-meaning company from trying to improve its safety program.
Although most unions do not block participation, some international unions
do not support the VPP (Catanzaro 1994).

Shortcomings of VPP

- Complex program that increases paperwork
  and safety and health program costs
- Increased government oversight/inefficient
  use of OSHA resources
- Is only for outstanding companies
- Statistics are skewed by good companies
- Must get union support
- Program needs formal codification to be
  viable in the future

Further, a 1998 article quoting Joe Anderson, Oil, Chemical and Atomic
Workers (OCAW) health and safety director, indicates that some union
members do not see the VPP the same way OSHA does. He states that the
"idea of an employer complying with OSHA only if it can do so voluntarily
makes me suspicious" (Karr, 1998). Apparently he thinks that participants
have a choice about regulatory compliance. However, it is clear that OSHA
expects each site to comply with every aspect of its regulations while in
the VPP, implement elements that are beyond regulation, and continuously
improve programs.

The VPPPA likes the program so much that its mantra is: *get it codified.*
It claims that the program, though effective, is "subject to budget cuts and
other pressures that may threaten its existence" because it is not a part of
the Occupational Safety Act. The program is providing enhanced protection
to "over 300,000 workers" and "is a critical component to ensuring increased
worker health and safety" (Edwardson 1999). The VPP is currently funded
through OSHA's Compliance Assistance program — a program that is less
than 1% of the agency's total budget. In spite of its small-budget status, the
VPP participants are not comfortable with possibly having a budget cut end
its existence. The VPPPA states that there is bipartisan support for a bill to
make the VPP permanent, and it is working toward a bill that will be
supported by all (Edwardson 1999).

Synopsis

The VPP program is very successful. It has certainly changed the typical regulatory agency/industry mode of operation for those in the program. Good will has definitely been developed — not only between OSHA and industry, but labor/management relationships have improved as a result of this initiative. Certainly, OSHA has made strides in getting better protection for workers since participants have an injury rate 55% below the norm for their industry.

Companies that have participated in the program enjoy the benefits of OSHA's advice; a nonconfrontational relationship; better business results in the form of lower worker compensation costs and injury costs; and employees who are at work rather than at home recovering from injuries suffered on the job. They also are able to fly the VPP flag in front of their factories and benefit from improved images with their communities and their employees. OSHA treats VPP companies as partners, expects them to help weaker companies, and even permits participants' employees to act in lieu of government employees in auditing their peers on VPP verification reviews.

The VPP program also helps industries by giving them a proven systematic way to manage safety and health. Companies are privy to best practices and industry standards through the mentoring program and can join the VPPPA, an association that will help them achieve the goals of the program. Facilities are at a lower risk of receiving enforcement action at their site when in the program. If violations are found during the verification visits, the company can correct the violation without receiving a citation or penalty.

Yet with all the positives of this program, it is difficult to see the ultimate benefit to the American worker. It is true that workers are being better protected at VPP sites, but *what about all of the other sites*? Since OSHA does not typically visit industrial sites very often, should not OSHA be focusing on the bad actors rather than the good performers? Wouldn't the public be better served if it concentrated on companies with real problems? The question of OSHA resource allocation is a legitimate one.

The VPP program is complex and costly to implement. Participation is limited only to companies that have better-than-average programs, and thus poor performers are not permitted into the program.

Some doubt exists whether the VPP has produced the stellar injury-prevention rates and cost savings — or whether the rates are good because only good companies are allowed in the program. Do the VPP companies have "from 60 to 80% fewer lost workday injuries than would be expected of an average site of the same size in their industries," as OSHA claims, because they are the best companies out there anyway? (OSHA 1999). Would these results occur if the VPP never existed? It also can be frustrating for those who want to be in the program but who are blocked from participation because their labor union does not want to participate. There is also a concern that the program could be eliminated by a budget cut, so there are some that are pushing to have the VPP codified.

The VPP has taught us several things about self-regulatory programs. First, it is possible to have excellent cooperation between government and industry. Government can trust industry, even to the extent that industry employees act as special government agents. Significant momentum can result from self-regulatory programs when industry gets behind them, as evidenced by the promotion of the VPP by the VPPPA. We learned that a self-regulatory program must not detract from achieving broad compliance of all industry if one of its purposes is to use government resources the most effective way possible. It is evident that the more inclusive we are with a scheme, the greater the benefit will be to society because more sites will be in compliance with the regulations.

It would be helpful to have as simple a program as possible. The VPP is not an easy-to-understand program, and there might be greater participation if it were less complicated. An additional idea is to reduce some of the more difficult and expensive requirements in order to make it easier for companies that want to do the right thing to participate.

It is clear that any self-regulatory program must have incentives for participation. Caution should be taken not to reward companies with increased regulatory scrutiny for their participation in a program while avoiding the poorer performing industries. The cooperative nature of the VPP shields the participant from violations noted during OSHA visits. Protection from violations is a very significant incentive for industry. It would therefore be wise to adopt this incentive in an environmental self-regulatory program.

Finally, a self-regulatory program must have some securities for its long-term viability. In the opinion of the VPPPA, there can only be assurance when the program is codified in regulation.

Bibliography

Catanzaro, Gerry and Weinberg, Judith (1994). Answers to some frequently asked questions on VPP. *Job Saf. & H.*, Summer, pp. 1, 25.

Edwardson, Antonia (1999). VPPA members visit Capitol Hill to encourage VPP legislation. *The Leader*, Available: http://www.vppa.org/Leader/Article.cfm?key ArticleID=73, June 23, 1999 (Spring).

Edwardson, Antonia (1999). What is codification? *The Leader*, Available: http:// www.vppa.org/Leader/Article.cfm?keyArticleID=74, June 23, (Spring), p. 1.

Fisher, A. (1997). Danger zone. *Fortune*, September, p. 167.

Garner, Charlotte A. and Horn, P. O. (1999). *How Smart Managers Improve Their Safety and Health Systems: Benchmarking with OSHA VPP Criteria*. Des Plaines, American Society of Safety Engineers, pp. 7, 21, 47, 50, 61, 67–69.

Gunningham, Neil and Johnstone, Richard (1999). *Regulating Workplace Safety*. Oxford, Oxford University Press, pp. 82–83.

Karr, A. (1998). Shooting for the Stars. *Saf. Health*, 157(6):74–79.

Minter, S. G. (1997). Dan Peterson: why safety is a people problem. *Occup. Hazards*, 59(1):39.

Mintz, Benjamin W. (1984). *OSHA History, Law, and Policy.* Washington, D.C., The Bureau of National Affairs, Inc., pp. 361–363.

OSHA (1996). *Revised Voluntary Protection Programs (VPP) Policies and Procedures Manual — 8.1a.* Available: http://www.osha-lc.gov/OshDoc/Directive_data/TED_ 8_1A.html, September 9, 1999 TED 8.1a(5/24/1996), pp. 14–15, 77–91, 106–117, 136–138, 140, 198–199.

OSHA (1999). The benefits of participating in VPP. Available: http://www.osha.gov/oshprogs/vpp/benefits.htm, September 23.

OSHA (1999). Myths about VPP. Available: http://www.osha.gov/oshprogs/vpp/myths.html, September 23.

OSHA (1999). VPP federal program growth. Available: http://www.osha.gov/oshprogs/vpp/fedgrow.html, September 23.

OSHA (1999). Why does OSHA need to come to my site? Available: http://www.osha.gov/oshprogs/vpp/original/onsite.html#Who, September 23.

OSHA (2000). Revisions to the voluntary protection programs to provide safe and healthful working conditions. *Federal Register,* 65:45649-45663:1–2, 22 (July 24).

OSHA (2001). Growth of VPP. Available: http://www.osha.gov/oshprogs/vpp/spgrowth.html, and http://www.osha.gov/oshprogs/vpp/fedgrowth.html, March 29.

OSHA (2001). An overview of VPP. Available: http://www.osha.gov/oshprogs/vpp/overview.html, March 29.

OSHA (2001). The benefits of participating in VPP. Available: http://www.osha.gov/oshprogs/vpp/overview.html, March 29.

Philips, J. (1999). International paper safety and health manager testifies on OSHA bill. *VPPPA News Updates.* Available: http://www.vpppa.org/News/PressArticle.cfm, June 23.

Phillips, Julie (1999). New ways of doing business with OSHA: OSHA reaches out to public and private groups to gain partnerships. *The Leader,* Available: http://www. vpppa.org/Leader/Article.cfm, June 23 (Winter), pp. 1–2.

Richardson, Margaret R. (1999). *Preparing for the Voluntary Protection Programs.* New York, John Wiley & Sons, pp. ix, 14, 16, 20–22, 29.

Ruhnka, John C. and Boerstler, Heidi (1998). Governmental incentives for corporate self-regulation. *J. Bus. Ethics,* 17(3311):313.

Sherrill, Leigh and Weinberg, Judith (1998). OSHA's VPP gets a little help from its friends. *Job Saf. & H.,* Summer, p. 31.

Siddiqi, Shahala and Johnson, D. (1999). Committed to excellence. *Ind. Saf. Hyg. News Mag.* Available: http://www.ishn.com/newsletter/9905/cover.htm, June 23, pp. 1–2, 5, 7.

Swartz, George (1995). OSHA's voluntary protection program: are you ready? *Prof. Saf.,* January 40:22–23.

VPPPA (2001). About VPPPA. Available: http://www.vpppa.org/About/index.cfm, April 2, p. 1.

Weinberg, Judith (1997). *OSHA's VPP Promotes Greater Worker Protection, Lower Costs,* NCCI, pp. 98, 100.

chapter five

StarTrack case study

Background

Many have wondered why the EPA has never developed a similar program to the OSHA Voluntary Protection Program (VPP). On the surface, the VPP seems to have all the elements that would be attractive to industry and government, especially more cooperation and better results. The closest environmental endeavor to the VPP was the self-regulation initiative developed by EPA Region I called StarTrack. Although the StarTrack experiment has ended, there are some attractive concepts that demonstrate the pitfalls and benefits of an environmental self-regulatory program. A further reason to study StarTrack is that it was used as the model for a nationwide program designed to motivate and reward top environmental performers called the National Environmental Performance Track (EPA Region I 2000). The Performance Track program was initiated in 2000, so there is sparse literature on this fledgling venture.

Since StarTrack had a 6-year history and there are scholarly works and data available, it is a good case to analyze. Our study covers the program history, eligibility for participation, program elements, and an examination of its benefits and criticisms. Finally, conclusions are drawn about what we learned from StarTrack about its use as a policy instrument.

> Publicly traded companies have their financial results audited by independent third-party accounting firms. Why not do the same in the environmental arena?

History

EPA's innovative StarTrack program was conceived by looking at an aspect of business and society seemingly unrelated to environmental protection: finance. Publicly traded companies have their financial results audited by independent third-party accounting firms in accordance with federal securities laws. Why not do the same in the environmental arena? This idea is even more attractive when considering the point made in Chapter 1: "over 100,000 facilities nationwide are subject to federal permits for air or water emissions or are RCRA large-quantity generators," and site inspections by regulatory agencies of these facilities are extremely low. In fact, they are so low that "less than 1% of these facilities" received multimedia inspections (air, water, and hazardous waste) over a 2-year period (EPA 1998). Since sites are not visited by agencies, there is no harm in exploring the possibility of having independent third-party environmental firms audit the compliance of the more responsible organizations.

StarTrack began as a cooperative effort. In 1994 the EPA published its Environmental Leadership Program (ELP) Pilot in the *Federal Register*. The Gillette Company had a pilot project in this program that evaluated the use of third parties to verify self-auditing programs and environmental management systems. Using this program as a model, EPA Region I and the New England states initiated StarTrack in May 1996 as part of a "regional leadership program." The Gillette ELP project audit guidance and the project agreements were used as the basis for the program (EPA 1998).

Eligibility

Unlike Project XL, which has a goal to experiment with various types of new ideas, StarTrack is a test of implementing a "third-party certification system that utilizes auditing and management systems to improve compliance and performance ... and provide facility performance information to the public and regulators" (EPA 1998). Participants may include any industrial facility, federal or state facility, municipality, or other organization with environmental management and compliance programs.

The application form in the StarTrack brochure consisted of only 12 questions. In order to participate, a company or site must have a good working relationship established with the EPA and state agency and have a compliance history that includes no "open enforcement actions or serious recent compliance problems." A formal compliance-auditing program should be in place, and management must be committed to EMS, pollution prevention, and continuous improvement.

To participate, a company or site must have:

- no open enforcement actions
- a formal compliance auditing program
- a commitment to EMS, pollution prevention, and continuous improvement

---

The StarTrack participant and the agencies sign a "Project Agreement and a Letter of Commitment" stating that the firm is committed to be in the program for 12 months. The EPA and/or the state environmental agency can terminate participation in the agreement with 30 days' written notice.

Since participants must already have a compliance-auditing program, a good compliance record, and a cooperative relationship with regulatory agencies, this initiative was only available to organizations that had mature environmental programs (EPA 1999). At the end of the program there were only 14 organizations participating in the StarTrack program. Other than the U.S. Coast Guard and the U.S. Postal Service, the participants were primarily manufacturing firms located in five different states (Connecticut, Maine, Massachusetts, New Hampshire, and Rhode Island). Most of the companies were large multinational firms such as Texas Instruments, International Paper Company, Spalding Sports, and GAF Materials (EPA Region I 2000).

Program elements

The main elements of StarTrack are:

- Annual regulatory compliance and Environmental Management System (EMS) self-audits
- Annual publicly available environmental performance report
- Third-party certification of the self-audit program every 3 years (EPA 1998)

This program is based on self-auditing and disclosure. An examination of these elements provides a clear understanding of what this experimental program involved.

Self-audits

The backbone of the StarTrack program is assurance of regulatory compliance accomplished by facilities auditing themselves rather than being audited by the EPA or state agencies. Annual auditing of a participant's regulatory compliance and EMS must be performed.

The program managers have developed several very detailed guidance documents that help StarTrack companies meet the goals of the program. The documents spell out every aspect of the annual audits, from auditor qualifications to inspection procedures to report protocols. The regulatory audit is a multimedia, facility-wide assessment. Traditionally, state or federal environmental agency inspectors visit a site based on a single medium. For example, if a site has a wastewater treatment plant, an inspector from the water resources group performs the audit. Corporations that have internal audit programs and focus on air, water, waste, and other issues typically perform multimedia inspections. The designers of StarTrack saw the merits of this holistic approach and embraced it.

StarTrack auditors must meet professional, objectivity, and independence requirements. First, they must have the skills necessary to perform a diligent regulatory and EMS review. Knowledge and training in federal, state, and local environmental regulations and industrial processes and pollution control are musts (EPA 1998). Also, they must be experienced with multimedia compliance audits, have at least "twenty equivalent work days of auditing and have participated on a minimum of four EMS" audits (EPA 1998).

Auditors can be from the same organization as the one undergoing audit provided that they report to a different unit than the one under audit and are not accountable to those responsible for the information investigated. It is not uncommon for large corporations to ask an individual or a team from the corporate staff to perform an environmental audit of a manufacturing plant. Typically, the corporate staff does not have the same manager as the manufacturing plant, nor does it eventually report to the same upper-management executive.

Auditors from outside the organization, such as environmental consultants, may be used if they do not have any conflict of interest. Examples of unacceptable relationships include history of contractual agreements with the audited site (other than previous environmental audits) and a direct financial stake in the outcome of the audit. All auditors must sign certifications that become part of the audit record stating that they have the proper environmental technical skills and can perform the review with objectivity and independence (EPA 1998).

The EPA describes exactly what it would like to see happen during the audit. Audit protocols must be developed that address applicable environmental regulatory requirements, EMS, and any beyond-compliance activities such as pollution prevention. The audit should be conducted during a time when the facility is in normal operations. If there is more than a one-shift operation, the auditor must make provisions to visit the site during each shift. There is a whole host of documents that must be reviewed. Some of the records to be reviewed include process flow diagrams, sewer maps, storage tanks, spill control and emergency response plans, training records, waste-generation records, permits, facility guidelines, corporate guidelines, production records, EMS program description, environmental policy and objectives, and targets to address all of the organization's significant environmental aspects. This limited list shows that the review is extremely thorough.

Each audit must include a facility inspection that incorporates visual observations, interviews of facility personnel, and, of course, document reviews. The EPA expects the inspection to cover all areas that have environmental significance, including production processes, waste-management areas, wastewater and storm water-control facilities, storage tanks, pollution-control equipment, and remediation activities. Sampling or testing is suggested if there is good reason, such as discoloration of soil, presence of unknown material, or presence of disallowed discharges. The EPA also addresses the potential use of photographic and video equipment to document any potential noncompliance areas.

The auditor should also "identify facility risks that if left unchecked may result in environmental/human harm" that are not specified in any regulation. An example is a PCB transformer located in an area where forklift traffic might damage it and cause a leak (EPA 1998). The EPA suggests the ISO draft Audit Guidelines as appropriate EMS audit methods. These guidelines are essentially ISO 14001 management system requirements (EPA 1998).

The on-site audit must include an opening conference to discuss the scope of the review with all the facility staff, and a closing conference to discuss all potential deficiencies. Any serious deficiencies should be corrected immediately if at all possible, and all deficiencies should be addressed as soon as possible. A final compliance audit report must be submitted within 30 days of the audit. The report should address "the auditor's observations, findings and recommendations." The compliance report is available for review at the StarTrack participant's facility. The EMS audit report must be sent directly to the EPA and state and local agencies as well as the facility within 30 days. In addition to the full audit report, a Compliance Audit Summary Report containing potential or actual noncompliances found must be sent to the regulatory agencies. The recommended contents of a report include:

- Introduction
- Audit procedures
- Facility background
- Audit results
- Conformance to audit guidance and protocols
- Certifications of auditor's independence, skills, and knowledge

Corrective action plans must be developed for all areas of noncompliance. At a minimum the plans should contain a detailed description of each noncompliance and the actions to be taken to fix the problem. The root cause of each problem and a description of how the site will prevent recurrence must also be discussed (EPA 1998). An example of the format for reporting corrective action from the StarTrack guidance documents is shown in Table 5.1 and demonstrates the level of detail required.

Annual performance report

The EPA states that the goal of the annual report requirement is to "provide relevant information on a company's environmental performance to EPA and the Public." The EPA's guide on annual environmental performance reporting prescribes a minimum set of reporting criteria. The agency encourages facilities not to restrict their reporting to this, but to report as much information as possible to demonstrate their facility's environmental performance. There are seven core indicators that must be included.

1. Nonhazardous solid waste generation
2. Hazardous waste generation
3. Water use
4. Energy use
5. Wastewater discharged

6. Air emissions (significant pollutants reported as part of the Toxic Release Inventory or greenhouse gases and ozone-depleting chemicals)
7. Significant environmental incidents (reported spills, noncompliance events, any enforcement actions)

These data are mostly reported as pounds per month or gallons per day. Facilities are encouraged to index their emissions to production units and to show trends over periods of years. Sites are also encouraged to report on environmental indicators other than pounds of emissions such as objectives and targets for significant environmental issues identified from their EMS and voluntary programs — i.e., community outreach programs (EPA 1998).

The EPA has a suggested reporting format demonstrated by one of the StarTrack participants, EG&G Optoelectronics of Salem, Massachusetts. The report has the following information:

Summary
A brief description of the company's commitment to excellence in environmental management, its ISO 14001, and partnership with EPA–New England with the StarTrack initiative.

Environmental performance trend data — Direct effects
Using 1992 as a base year, the company lists various graphs indicating downward trends in hazardous waste generation, air emissions (volatile organic compound emissions), electrical consumption, water consumption, and solid waste, with a short summary describing its reduction programs for each of these areas.

For example, the chart shown in Figure 5.1, as well as several others for the seven core indicators, appeared in the EG&G report. The report stated that recycling and reuse of packaging material resulted in this outstanding reduction in solid waste generation. The other areas mentioned were reported in the same way, and they also demonstrated impressive downward trends in emissions.

Environmental performance trend data — Indirect effects
This section addresses aspects of EG&G's environmental program that are not quantified by hard data — complaints received and how they were addressed, training of employees, and a description of a program that grades suppliers on ISO 14001 conformance.

Auditing and corrective action program
This consists of a description of the quarterly internal audit process the company developed for its EMS, ISO 14001, and regulatory requirements. Also included are the annual audit of its ISO 14001 by certification body, the National Standards Authority of Ireland, internal corporate group audits, and

Table 5.1 Suggested Format for Tabulation of Corrective Action Plans

Issue description	Issue type (regulatory, management system, company policy)	Regulatory, permit or policy citation	Issue status	Corrective action	Scheduled completion date	Actual completion date	Preventive action (management system changes)	Scheduled completion date

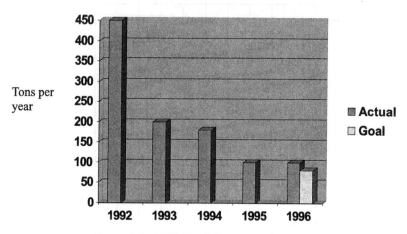

Figure 5.1   EG&G solid waste reduction.

the third-party audit performed by an environmental consultant. Note that the National Standards Authority of Ireland is also an independent third party that is approved to verify conformance to ISO 14001 and issue registrations to companies certifying that they meet the management standard. The audit report and corrective actions are attached to the annual report.

Management review

Management's interaction with the company's EMS is described. EG&G has monthly, bimonthly, or quarterly reviews that address "all objectives and targets of the EMS." An annual Environmental Management Plan is prepared that includes a summary of the year's objectives and targets, costs, material balances sheet, hazardous waste, and resource consumption.

EPA — New England's environmental leadership program and StarTrack initiative

This section describes the company's involvement with the StarTrack program, internal corporate audits, third-party review of audits, and ISO 14001 and Irish Standard 310 certifications (EPA 1998).

The EPA posted the annual reports on its Web site. Therefore, all information submitted was easily be accessed by the public (EPA Region I 2001).

Third-party certification

The purpose of third-party certification is to have an independent objective evaluation and report of the facility's overall program similar to an accounting firm's evaluation of a publicly traded corporation's financial status. Third-party assessment is independent testing and verification of a site's compliance and EMS audits and findings. It also serves the purpose of preparing an EMS improvement plan that addresses the findings of the certifier's report.

> The purpose of third-party certification is to have an independent, objective evaluation and report — similar to an accounting firm's financial status evaluation of a publicly traded company.

The certifier must be an independent third-party contractor such as an environmental consultant or ISO registrar. He or she must not have any conflict of interest with the site being evaluated and must be in a position to give an independent evaluation. He or she must have at least the same knowledge and training that is required for the self-audits. A certification statement is also required for the auditor to verify that he or she has the proper credentials to perform the review and has no conflict of interest that can skew judgment (EPA 1997).

The certifier must submit a proposed work plan to the StarTrack facility and the agency, documenting the approach to be used in performing the review. Any comments that either party has must be submitted in writing to the certifier, who must in turn send responses to each party. The actual review covers the observation of the site's compliance and EMS audit programs and includes "independent observation and testing ... to verify the validity of audit findings." The certifier must attend and observe the StarTrack site's annual self-compliance and EMS audit program. The evaluation must:

- Determine if the StarTrack Guidance Documents are adhered to (e.g., auditor qualifications, self-audit program)
- Assess the site's regulatory compliance and EMS to determine if it can prevent and detect noncompliance, meet its beyond-compliance commitments, and manage changes to regulatory requirements
- Assess the site's regulatory compliance status
- Determine if beyond-compliance activities are consistent with goals and objectives

The certification review results in a report that either recommends corrective actions to address deficiencies or a statement that corrective action is not necessary. A certification statement must document the status of the site's regulatory compliance, beyond-compliance activities, and EMS. The suggested headings of the certification review report are listed below.

Each heading has specific requirements the certifier must check. Two of the more far-reaching elements are the assessment of the site's EMS to ISO 14001 management systems and the ability of the facility to foster communications with its corporate headquarters regarding environmental

---

Certification Review Report Elements
- Introduction
- Audit procedures
- Facility background information
- Assessment of environmental management systems
- Assessment of compliance status and compliance management system

---

compliance. Also suggested are critiques of the StarTrack facility's environmental staff's education and training, capability for compliance with federal, state, and local regulations, and an evaluation of the corrective action program (EPA 1997). These requirements have the look and feel of an internal corporate audit.

Finally, there must be a certification statement. There is a model statement that the certifier must complete. It attests that the company provided all necessary information and is or is not following the StarTrack guidelines. A decision must be made whether the company receives full, conditional, or non-certification. Full certification means the site meets all or substantially all the requirements. Conditional certification means that the site generally meets the requirements but there are deficiencies that are not deemed significant. Conditional facilities must have the third-party assessment annually until they achieve full certification status. Non-certification means that the site requires significant improvement and has inadequate compliance or EMS audits. A site in this category is terminated from the program (EPA 1997).

The certification review report must be sent to the agency and the StarTrack facility directly without being evaluated first, unless drafts are submitted simultaneously to both parties. An EMS improvement plan must be submitted to the agency with the certifier documenting actions the site will take to correct any deficiencies. The certifier must perform a follow-up visit within 60 days after the EMS improvement plan is submitted. Then another report is issued to the agency indicating if the facility has addressed the deficiencies noted.

Enforcement action will not be pursued for violations disclosed in the StarTrack program unless they are criminal, they result in serious actual harm, or will result in or are likely to result in imminent and substantial endangerment or damage to the environment.

The EPA placed a disclaimer in its guidance document to make it clear that it "shall not be bound by any determination by the StarTrack Facility, the Compliance Auditor(s), or the Certifier that the StarTrack Facility has achieved compliance" (EPA 1997). This means that the StarTrack program does not operate in lieu of the EPA and state environmental agencies in the matter of determining if a site is in compliance with regulations. All compliance and EMS audit reports must be submitted to the agency. Any violations found must be corrected within 60 days, unless it is not feasible to do so. In such cases a written waiver can be obtained to go beyond the 60-day time limit. Enforcement action will not be pursued for violations disclosed in the StarTrack program unless they are criminal; they result in serious actual harm or will result in or are likely to result in imminent and substantial endangerment or damage to the environment or to public health; they result in significant economic benefit; they are violations of judicial or administrative order or consent agreement; or they are repeat violations as defined in the EPA's Audit Policy (EPA 1998).

Analysis of StarTrack

Benefits of StarTrack

The StarTrack program was an innovative attempt at achieving compliance with regulations and increased environmental performance through the use of third-party certification and management systems. EPA Region I *should be commended* for its bold attempt at framing a cooperative program that has the potential to reduce compliance oversight costs and help focus agency resources while rewarding proactive organizations. EPA Region I has produced two excellent documents that give a balanced evaluation of the program: the *StarTrack Year One Final Report* and *The National Expansion of StarTrack Report*. The Year One Report is an excellent source of data, reports, and surveys of StarTrack participants and other stakeholders. In reading the report, an obvious benefit that surfaces is increased compliance with regulations.

> EPA Region I should be commended for its bold attempt at framing a cooperative program that has the potential to reduce compliance oversight costs and help focus agency resources while rewarding proactive organizations.

The program requires all participants to perform annual compliance audits of their operations and to disclose the findings to the EPA and state environmental agencies. The fact that these reviews are multimedia audits and occur on an annual basis is a plus. As noted previously, the EPA's own statistics indicate that *less than 1% of all regulated facilities receive multimedia inspection.*

There is evidence that the program is achieving its regulatory compliance potential when evaluating some of the findings that were disclosed to the EPA by their StarTrack partners. The *StarTrack Year One Final Report* includes a copy of International Paper's (IP) environmental audit report dated

February 25, 1997. This report is the result of a 5-day audit of the company's Androscoggin paper mill in Jay, Maine, by IP's environmental auditing group that was witnessed by members of the EPA, Maine DEP, and the Town of Jay. The environmental consulting firm ERM–New England, Inc. witnessed the audit as the third-party reviewer. Besides a very detailed description of the mill's operation, which is not typically made available to the public, there is a list of deficiencies with regulatory and corporate requirements. It is obvious from this list that the audit was not just a superficial review. Significant noncompliance issues were raised; examples include not operating a thermal oxidizer air pollutant control device above the 1500°F temperature required by a Maine air pollutant permit and several nonconformances with the site's Spill Prevention Countermeasures and Contingency (SPCC) plan (lack of storage tank inspections) (Kraske 1997).

Significant noncompliance issues were raised; all regulatory issues except those associated with the facility's SPCC plan were corrected approximately 3 weeks after the CAP was completed.

The fact that these findings surfaced is impressive. Even more convincing is the fact that IP produced a corrective action plan (CAP) to address each gap, and "all regulatory issues except those associated with the facility's SPCC plan were corrected approximately 3 weeks after the CAP was completed" (Anonymous 1999). Similar issues were identified at Spalding Sports Worldwide at its Chicopee, Massachusetts manufacturing plant during its compliance audit. Violations of PCB Control Plans, SPCC Plans, and lack of submitting a Massachusetts Contingency Plan report were cited and corrected (Sweetman 1997).

Texas Instruments (TI) has also used the StarTrack system to identify and correct regulatory problems. According to its 1998 Report on the Environment for the Attleboro-Mansfield, Massachusetts facility, its third-party audit resulted in the discovery of excessive wastewater discharge (cyanide at 0.31 mg/l compared to the permit limit of 0.18 mg/l) without promptly notifying the agency. Although TI did notify the agency, it did not do so within the specified 24-hour limit. Other issues identified included failure to submit an air emissions report and failure to complete an annual storm water assessment (Veale 1998).

Further evidence of improved compliance was documented in a research paper performed on behalf of the National Academy of Public Administration, an independent organization chartered by Congress to improve governance. The report documents a statement from an EPA Office of Enforcement and Compliance Assurance (OECA) employee stating that "there is no evidence to suggest that environmental performance at StarTrack companies is declining, or that companies are taking advantage of the penalty-mitigation provisions offered by the program." The EPA can take enforcement action on program participants if there are egregious violations or if disclosed violations are not corrected within 60 days. There were no enforcement actions taken at StarTrack sites (Nash 2000).

Based on the required annual reports made available through EPA Region I, companies that participate in the StarTrack program demonstrate good environmental performance. Already mentioned was the performance of EG&G, which indicated a trend over several years of significant reductions in pollutants. Other examples include Spalding's Performance Report that indicated significant reductions indexed to dozens of golf balls produced. Wastewater generation was reduced from 10.15 to 1.09 gal/dozen from 1990–1996, VOC emissions reduced from about 1.52 to 0.68 lb/dozen from 1992–1996, and the amount of materials recycled remained about the same from 1992–1996 (~800,000 lb/year) (Sweetman 1997).

> The public has access to information that is not available in any other forum.

Likewise, Texas Instruments reduced its water use 11% from 1997 to 1998. Almost all 17 of the toxic chemical releases TI is required to report to the EPA and MA DEP were reduced between 1996 and 1998, and similar reductions have occurred for volatile organic compound emissions, heavy metal discharges, and hazardous waste (Veale 1998).

Besides the environmental benefit of reductions in emissions, which are not mandated by any environmental regulation or statute, the public has access to information that is not available in any other forum. Participants' annual performance reports indicating the amount of pollutants emitted and other programs are publicly available (EPA 1998). This information is even available on the Internet. Texas Instruments posted its 35-page Annual Environmental Performance Report on its Internet site. This report not only addresses the good things TI is doing; it also indicates the regulatory non-conformances that were found during annual audits as part of the StarTrack program (Veale 1998).

> A majority of the companies participating in the program do so for the recognition.

It appears that a majority of the companies participating in the program do so for the recognition that results. The page discussing participation in StarTrack in Texas Instruments' annual report is laced with stars and statements like "beyond compliance" and "the EPA has distinguished TI Attleboro as an environmental leader" (Veale 1998).

There is no doubt that public perception as an environmental leader is good for community relations. International Paper received the Governor's Award for Environmental Excellence in the environmental leader category.

The award acknowledged the participation in the StarTrack program. It is hard to get the kind of publicity that has the governor of a state proclaiming that a company has "demonstrated initiative, inventiveness and willingness to take risk to protect the environment" and goes "beyond compliance" (Perry 1999).

The EPA, to its credit, went out of its way to recognize companies in the program. It stages events to hail its partners and makes press releases available to the public and StarTrack participants. Participants easily obtain notable quotes. For example, this from John P. DeVillars, former EPA New England Administrator, stating:

> Companies like BOC Gases and International Paper Co. are doing exemplary work developing and monitoring their own environmental systems ... that's good for the environment and good for taxpayers because it allows EPA to channel its energies to other, less environmentally responsible companies (EPA 1999).

EPA's attempts at giving credit to participants has been one of the shifts in government/industry relations that comes from cooperative programs. David Guest, EPA coordinator of the StarTrack program, states that companies like the program because it is "more of a partnership than an adversarial association" (St. Ours 1998). The development of this program has purposefully been a partnership between industry and government. It began with Gillette's environmental leadership project, and industry has been a development partner since then. Firms are asked to give their opinions and are polled for feedback on the effectiveness of the program.

The development of this program has purposefully been a partnership between industry and government.

Audits have been attended by EPA and state regulatory agency staff. In some cases the regulatory agency observers suggested improvements to audit programs and EMS. In the case of EG&G, the agency was given access to observe an ISO 14001 audit by the company's registrar. EG&G, when asked, also "provided case-study presentations at some EPA internal ISO 14001 training sessions and outreach sessions to the benefit of agency personnel." (EPA 1998). In cases like this, cooperation results in trust and better environmental performance because both the company and the agency are learning from each other. ·

Environmental groups have also benefited from the partnership atmosphere of StarTrack. The EPA encourages the participation of environmental groups, although this has not been made a requirement of the program. On at least two occasions environmental groups were able to participate in audits. The Massachusetts Audubon Society received an unprecedented opportunity for an interest group — attending a Texas Instruments audit. In a letter to the

EPA, Louis J. Wagner, a water resources specialist with the society, states that the program is a sound approach because "EPA and the state environmental agencies will never have sufficient staff and budget to do complete reviews of compliance at all industrial facilities on a regular basis." If the program will allow resources to be focused on industries with poor compliance records, it will "prove a very beneficial program" (Wagner 1996).

A National Academy of Public Administration research team interviewed members of the community and environmental groups that attended StarTrack audits. Two participants from environmental groups agreed that facility operators were "very positive" with their involvement. They were very upbeat about the program and felt that it "provided for non-confrontational relationships to develop." Members of local government that observed the audits were also happy with the initiative. One said that the company was "tough on themselves." Attendance on the audit was a "turning point" for one observer; he thought that the experience helped change his opinion about the site. It allowed him to think that a "more cooperative relationship could yield greater environmental improvement than nit-picky enforcement" (Nash 2000).

One of the benefits of being a StarTrack company is enforcement discretion. The discretion comes in the form of "mitigated penalties for non-egregious violations" (PCSD 1999). Violations that are not of a serious nature are permitted as long as they are corrected within 60 days of the date of discovery. Another part of the discretion is the potential for the reduction of inspection priority. However, the National Expansion of StarTrack report states that questionnaires regarding incentives indicated that reduced inspections and penalty amnesty were not an important incentive (Hale 1998). This doesn't seem to correlate with some of the written comments received from participants that were surveyed by the EPA. Answers to the question regarding to what extent regulators should rely on audits in making decisions regarding inspection priority included:

- Participants should go to bottom of the watch list.
- Compliance inspections could be reduced to a very low priority.
- If a company is trying to do the right thing and has systems in place to find and correct and prevent nonconformance, then reduced inspections and other benefits are warranted (Holbrook 1997).

Therefore, it appears that the amnesty and inspection priority issues are important benefits to StarTrack companies.

Benefits of StarTrack

- Improved environmental compliance and performance
- Increased publicly available information
- Improved public image/notoriety for the participant
- Develops good relationships with regulators/industry/community
- Penalty mitigation and potential decreased inspections

Criticism of StarTrack

Although there are many good things about StarTrack, there are several opportunities for improvement. First, my opinion of the *StarTrack Year One Report* is that it was a good evaluation of the program. There is indication that the EPA has taken steps to correct some of the problems noted, such as developing more specific program requirements via the StarTrack guidance documents. Nevertheless, there is still room for improvement.

The most frequently cited criticism is that the benefits of participation are not clear.

There were nine participants when the program was initiated (Kuhn 1998). As of October 1999, 3 years later, there were 12 (EPA 1999); by the end of the program there were only 14 participants (EPA Region I 2000). Of the original nine, one company was denied readmission because of compliance problems. In 1998 EPA sent out over 1300 letters to the top toxic chemical emitters in New England soliciting participants for the StarTrack program (Hale 1998). Despite this effort, there was not much interest in joining the program.

This low rate of participation points to the most frequently cited criticism: the *benefits of participation are not clear*. A company that decides to enter the program goes out on a limb. It must make internal audit and environmental performance reports available to the public. This type of disclosure is not required by any environmental statute; therefore the company is making more data available to the public. Granted, many large corporations publish environmental performance data, but this type of information is typically aggregated for their worldwide operations. The public reporting function of the StarTrack program requires facility-specific performance data. Companies are generally uncomfortable releasing this type of information because environmental groups could possibly use the data against them. Additionally, the participant must pay for two annual audits and a third-party certification every 3 years.

The benefits promised to participants were:

1. Partnerships
2. Penalty mitigation
3. Inspection relief
4. Rapid processing of permits
5. Recognition

Benefits other than recognition (in which EPA did a very good job) and penalty mitigation never really materialized. Some sites were still routinely inspected, and permit applications were not handled any differently. Partnerships were forged; however, because routine inspections still occurred despite the participant releasing a host of environmental performance information and performing many inspections, it cannot be concluded that real partnerships occurred. The National Academy of Public Administration team concluded that the benefit delivered — recognition — will not on its own "move facilities in the direction of excellence" (Nash 2000).

> Because routine inspections still occurred despite the participant releasing a host of environmental performance information and performing many inspections, it cannot be concluded that real partnerships occurred.

In the *StarTrack Year One Final Report*, Spalding comments that "continued participation in StarTrack could be further enhanced if the benefits were clearly defined, expanded and utilized." The EPA even states in the same report that "direct costs of participation in the pilot generally outweigh current benefits" (EPA 1998). Costs of bringing in a third-party certifier can range from $10,000 to $30,000, and this does not take into consideration the costs associated with increased reporting and annual self-audits (Kuhn 1998).

Cary Coglianese, of Harvard's John F. Kennedy School of Government, and Jennifer Nash wrote that companies believed the "EPA was not able to provide the benefits — public recognition, fast-track permitting, and inspection relief — that it had promised." Therefore, participation in the program remained low (Coglianese 2001).

In a meeting hosted by the EPA with the Connecticut Department of Environmental Protection (DEP) and environmental groups, the DEP claimed that it struggled with assessing the benefits, and it sees other companies also "struggling to define what's in it for them." The environmental advocacy group Connecticut Fund for the Environment also brought up an excellent point regarding program benefits: "What resources are saved?" (Snyder 1996). One of the main drivers of the program is to outsource some of the compliance inspections via third-party certification and focus on the poorly performing companies. As the program goes forward, the agency will have to show clearly that the program does not use a lot of agency resources and helps focus resources to get the biggest benefit for the environment.

> The environmental advocacy group Connecticut Fund for the Environment also brought up an excellent point regarding program benefits: What resources are saved?

Further stating this concern is Louis Wagner of the Audubon Society: "EPA and state agencies must make it very clear that any savings associated with StarTrack will result in the redirection of compliance monitoring and enforcement efforts, not a reduction in resources directed to these activities."

He agrees that it is not clear how the program benefits participants (Wagner 1996). Besides the benefits of recognition, partnership, and penalty mitigation for minor violations, there is little evidence of the other promised benefits. The StarTrack brochure and the EPA's Web site promise modified inspection priority and "Express Lane service for permits and other regulatory actions," but these benefits never materialized (EPA 1999).

> The program has too much bureaucracy, and this adds to the confusion. An annual compliance audit that results in a compliance audit report, compliance audit summary report, and a corrective action plan report are required.

It is clear that the EPA has tried to put more processes and procedures into the program as a result of comments from the Year One Report and subsequent guidance documents. Some of the issues that were raised were the qualifications of auditors and the independence that they must have in order to give third-party certification of the program. In most cases it appeared that the participants' corporate headquarters environmental staff conducted the annual compliance and EMS audits, and environmental consultants did the third-party certification. On one occasion, however, the same consultant did the annual assessments and the third-party certification (EPA 1998). This was probably the impetus for clarifying the requirements for the independence of auditors in guidance documents.

The program has *too much bureaucracy*, and this adds to the confusion. An annual compliance audit that results in a compliance audit report, compliance audit summary report, and a corrective action plan report are required. The EMS must also be audited annually, and there is a separate guidance document for this. It requires an annual EMS audit report and EMS implementation plan for addressing deficiencies. The participants are also required to submit an annual performance report indicating amounts of emissions and significant environmental incidents. The third-party certification audit must occur every 3 years and results in a certification review report, certification statement, an EMS improvement plan for deficiencies, a follow-up visit by the certifier, and another report on the status of corrections. All of these steps have procedures and guidelines.

There is definitely room for streamlining the process and standardizing the system. One example would be to provide regulatory agency checklists for the auditors so everyone is checking for the same things. Another reasonable step would be to allow ISO 14001 registrars (independent firms recognized by an accreditation organization to issue ISO 14001 certificates) to do the EMS audits without any additional reporting or documents. If a site were ISO 14001 certified, any further auditing would be redundant with the StarTrack EMS requirement because ISO 14001 has requirements for reporting, verification, and auditor qualifications.

Granted, more resources are used when developing a new program than once it is fully rolled out. Nevertheless, the EPA's goal was to move to a more self-regulatory compliance system that would require few resources to oversee the top performers so the agency could focus on less stellar sites. The bureaucracy did not extend only to the reports and audits required by the participants, but also to the agency. Besides the time needed to review all the reports, state and federal environmental staff members routinely observed the third-party audits (Nash, 2000). Therefore, there were more resources extended on firms that were voluntarily performing audits. If a site in a voluntary compliance program is going to be reviewed as frequently as it would have been if it were not in the program, *what's the point?* Shouldn't more focus be placed on the poor performers?

If a site in a voluntary compliance program is going to be reviewed as frequently as it would have been if it were not in the program, what's the point? Shouldn't more focus be placed on the poor performers?

Some very good ideas that could simplify the whole process were mentioned in the *National Expansion of StarTrack* report. One is to utilize professionals already certified by an existing environmental organization. Some certifications given by organizations such as the Board of Environmental Auditor Certifications (BEAC) and Certified Environmental Auditor, if deemed stringent enough, could rid the EPA of the worry over auditor quality and ethics (Hale, 1998). Some state regulatory agency staffs have questioned "the ability of StarTrack's third-party certification system to be as effective as regulatory agency enforcement inspections" (Pendleton, 1999). Utilizing existing certification groups, assuming their programs are tough enough, would give these state agency personnel more comfort.

There is also the question of whether third-party certification is legal. The EPA claims that it cannot delegate any "governmental functions to parties outside the agency." The state environmental agencies and the EPA "must retain under current legal authorities, the final determination of a regulated entity's compliance status and what enforcement response, if any, is appropriate in any given instance of non-compliance" (EPA, 1998). This is the reason why the draft third-party certification guidance has a disclaimer stating that the agency is not "bound by any determination" that the Star-Track facility "has achieved compliance" (EPA, 1997).

Under the Toxic Substance Control Act (TSCA), a certification for individuals to mitigate the effects of lead-based paint was enabled by an amendment to TSCA known as the "Lead Based Paint Exposure Reduction Act." In like fashion, the Massachusetts Licensed Site Professional program allows nongovernmental professionals to oversee the cleanup of contaminated properties (Superfund sites). This was also made possible by an amendment to the state's Superfund law, M.G.L. 21E (Hale, 1998). At this point it doesn't appear that the EPA believes that a statutory change is necessary to enable StarTrack. Yet, if this program is to go national, the question of statutory amendments must be addressed.

Rigorous requirements that do not add any value also seem to detract from the program. Both the draft guidance for the compliance and EMS audits require the auditor(s) to be on site for all shifts of operation if the site operates on more than one shift. This is required regardless of whether the off shifts have any unique operations or characteristics. They also suggest that all production processes and operations be visited and advise the use of photographic or video-graphic documentation (EPA 1998). Other requirements are auditing facility programs that go beyond compliance — i.e., pollution prevention and waste reduction opportunities, and the submittal of far too many reports (EPA 1998).

Evidently, environmental groups were not involved in the program development process as much as they would have liked. The meeting with Connecticut environmental groups indicated that they felt there was not enough public input in the selection of the companies to be in the program and took issue with calling the participants environmental leaders (Snyder 1996). Similarly, the Massachusetts Audubon Society felt that if the program were going to expand, the public would "need to know that environmental interest groups support it" (Wagner 1996). It is prudent to try to partner with nongovernmental interest groups to solicit their input and support in order to gain the greatest public acceptance of the program.

Poor or middle-of-the-road companies would not have much reason to get involved due to the lack of benefits.

Another critique is whether the program actually attributed to the improvement of the performance of participants. A case can be made that the organizations involved are the top performers; poor or middle-of-the-road companies would not have much reason to get involved due to the lack of benefits. So the question becomes, since a requirement to join StarTrack requires a history of pollution prevention and an environmental management system that requires improved performance, wouldn't these sites improve their performance anyway? (Nash 2000).

Criticisms of StarTrack

- Overly complex
- High costs for industry to participate
- Did not reduce agency oversight of top performers and therefore did not help focus the agency on poor performers
- Not enough incentives to participate

Synopsis

After studying StarTrack in detail, what lessons have we learned from this self-regulatory initiative? There is no doubt that the program has increased regulatory scrutiny at industrial sites. The current mode of inspecting industrial facilities is to perform single-media inspections, which, according to the EPA, do not occur very often. The StarTrack approach is a *more holistic method* of looking at all of a site's environmental regulatory obligations on an annual basis. Based on the review of some of the reports submitted by participants, the annual audits picked up on some significant issues that were promptly corrected.

This experiment seems to have been successful in driving compliance and environmental improvements with a carrot rather than a stick.

In addition to improved regulatory compliance, the program has resulted in *beyond-compliance behavior* by requiring ISO 14001 management systems. This implementation of management systems in addition to forcing a more systematic way of evaluating environmental issues also requires continuous improvement. Similarly, more information is made public through the annual reporting requirement. This fact alone can drive improvement. No company wants to have the public see data indicating that its waste and air emissions are increasing over time.

The system also has the added benefit of third-party certification of compliance and EMS. This experiment seems to have been successful in driving compliance and environmental improvements with a carrot rather than a stick. The EPA has made sure that participants are recognized for their commitments through staged events and press releases. Companies are also afforded violation amnesty for nonserious breaches such as paperwork infractions.

There is an obvious air of *cooperation and trust* with the state agencies/EPA, the participants, and, in some cases, environmental groups. Having the StarTrack companies help develop the program and be asked for their input is a new way of ensuring that regulatory compliance and environmental improvements occur.

The EPA recognizes that it has to *work on the benefits of participation*. Some promised benefits, such as expedited permitting and modified inspection priorities, have not materialized. If EPA does not make this type of program more attractive to others, it will likely only get companies interested in the public relations value of being called a "Star" to participate. EPA also has to broaden its own internal benefits. One of the mantras of the StarTrack program is better use of resources to focus on poorly performing industries. If it is expending a lot of resources on the oversight of this program, the EPA is not obtaining the benefits on its end.

One of the ways to improve the benefits for all is to simplify and systematize as much of the program as possible. At StarTrack's inception, ISO 14001 was not well understood. However, now that the EPA and companies have more experience with this management system, it is an ideal way to simplify the management systems part of the program. EMS is a beyond-compliance activity; so reliance on the existing system of registrars, certification bodies, and auditors has no drawback to EPA. Why should it set its own standards and require its own reports for EMS when a reputable system already exists? Shouldn't the agency spend more effort on the compliance audits and third-party certification (or verification) of compliance so it can gain confidence in this aspect of the program? This should result in greater environmental protection because of better environmental compliance and more time to focus on the bad actors.

Further, the elimination of checks that are rarely performed during routine inspections by agency staff, such as visiting the site on third shift and auditing beyond-compliance activities, will also simplify the program. Based on the agency's experience with only reporting data (e.g., Toxics Release Inventory), the reporting alone drives the proper behavior — less pollutants into the environment.

Before a program like this is rolled out throughout the country, all potential regulatory snags should be removed. If an amendment to an Act or new legislation is necessary, then efforts should be made expedite this task. If the agency would like to receive confidence in this program from the public, it should try to obtain more environmental group input into the process. The more groups that see the value of a program like StarTrack, the more effective it will be.

StarTrack has taught us a lot about self-regulatory programs. Third-party certification of compliance definitely has some promise. In the words of a Connecticut DEP staffer, "facilitating inspections without DEP staff is a tantalizing concept" (Snyder 1996). If StarTrack can prove that agency resources can be realigned away from good actors and to the bad, it can go a long way to help us do more with current resources.

In the words of a Connecticut DEP staffer, facilitating inspections without DEP staff is a tantalizing concept.

Third-party assessors can truly identify significant compliance issues that would most likely not have been efficiently picked up by agency inspectors due to the multimedia approach of the self-audits. Further, there is no downside or risk to state agencies or the EPA if they reserve their right to inspect the company and do not hold the third-party certification as absolute proof the site is in compliance. The bottom line is they do not have the staff to visit sites nearly as often as they would like, so any form of compliance checks beyond their own is a plus.

We have seen that the benefits of such a program have appeal to industry. It is obvious that companies like to be called a Star by environmental agencies, and this alone can attract some. Companies also like amnesty for self-identified noncompliance issues. Further, the concept of reduced inspection priority seems to be a good idea. Therefore, significant benefits are going to be necessary to attract as many companies as possible into a self-regulatory program to increase the critical mass of sites in full compliance with the regulations.

Bibliography

Anonymous (1999). StarTrack recognizes class leaders in EMS. *Environ. Manager's Compliance Advis.,* April 19, No. 493, p. 11.

Coglianese C. and Nash, J. (Eds.), *Regulating from the Inside.* Resources for the Future, Washington, D.C.

EPA, Region I (1997). StarTrack Program Guidance Document, *Draft Certification and Facility EMS Improvement Plan Guidance.* Boston, Environmental Protection Agency, Region I, pp. 1–9, 11–12.

EPA, Region I (1998). StarTrack Program Guidance Document, Appendix I, 1997. *Draft Guidance for Compliance Audit, Compliance Audit Report and Facility Corrective Action Plan.* Boston, Environmental Protection Agency, Region I, pp. 2, 5–8, 8–14.

EPA, Region I (1998). StarTrack program guidance document, Appendix II, 1997. *Draft Guidance for Environmental Management Systems Audit, Environmental Management Systems Audit Report and EMS Implementation Plan.* Boston, Environmental Protection Agency, Region I, pp. 1–9, Appendix IIA, 1–23.

EPA, Region I (1998). StarTrack Program Guidance Document, *Draft Guidance for Annual Environmental Performance Report.* Boston, Environmental Protection Agency, Region I, pp. i–ii, 1–3, 22.

EPA, Region I (1998). *StarTrack Year One Final Report.* Boston, Environmental Protection Agency, pp. ii, 1–2, 5–11.

EPA, Region I (1999). Better environmental performance through environmental management systems and third party certification. Available: http://www.epa.gov/region01/steward/stack/overview.html, October 28, p. 1.

EPA, Region I (1999). Current roster of StarTrack participants. Available: http://www.epa.gov/region01/steward/stack/roster.html, October 28, p. 1.

EPA, Region I (1999). EPA recognizes two Maine facilities for their work to become environmental leaders. Available: http://www.epa.gov/region01/pr/files/092299d.html, October 28, p. 1.

EPA, Region I (2000). EPA New England's StarTrack is model for national environmental achievement program. *Environ. News,* Release 00-07-03, p. 1.

EPA, Region I (2001). StarTrack facility environmental performance reports. Available: http://www.epa.gov/region01/steward/strack/epr.html, March 29, p. 1.

Hale, Rhea (1998). *The National Expansion of StarTrack.* Boston, U.S. Environmental Protection Agency, Region I New England, pp. 8, 17–18.

Holbrook, Jean E. (1997). *StarTrack Program Evaluation.* Boston, United States Environmental Protection Agency, Region I, p. 11.

Kraske, Charles R. (1997). *Summary Environmental Audit Report for the Androscoggin Mill.* Jay, International Paper, pp. 1–11.

Kuhn, Lauren, Langer, Jenn and Pfeiffer, Amy (1998). *Designing a Provisional System for StarTrack: An Environmental Management Strategy for the U.S. Environmental Protection Agency.* Boston, Massachusetts Institute of Technology, pp. 3, 11.

Nash, J., Ehrenfeld, J., MacDonagh-Dumler, P., and Thorens (2000). *ISO 14001 and EPA's Region I's StarTrack Program*. National Academy of Public Administration Research Paper Number 2 (June), pp. 5–9, 11, 12, 60, 61–63.

PCSD (1999). *President's Council for Sustainable Development Environmental Task Force Review of the StarTrack Program*. Washington, D.C., U.S. Environmental Protection Agency, New England, Region I, p. 2.

Pendleton, Susan (1999). Improving environmental performance: EPA's StarTrack program. *New England's Environment* (June), p. 2.

Perry, Donna (1999). Governor honors IP's environmental leadership. Lewiston, MA, *Sun–Journal*, p. 36.

Snyder, Gina (1996). *Meeting with Connecticut Environmental Advocacy Groups*. Boston, U.S. Environmental Protection Agency, Region I, pp. 1–3.

St. Ours, Denise (1998). StarTrack: an environmental regulatory framework for the next generation? *Int. Environ. Sys. Update* (November), p. 12.

Sweetman, William (1997). *StarTrack Environmental Performance Report for 1996*. Chicopee, MA, Spalding Sports Worldwide, pp. 8, 11–13.

Veale Jr., Francis (1998). *Report on the Environment: A Summary of Environmental Performance*. Attleboro, Texas Instruments Inc., pp. 1–35.

Wagner, Louis J. (1996). *StarTrack Program*. Boston, Massachusetts Audubon Society, pp. 1–2.

chapter six

Benchmarking industry voluntary initiatives

Background: The greening of industry

The greening of industry is a movement that began in the 1980s and has been growing ever since. One of the driving forces for this movement was the weight of media scrutiny and public opinion in the wake of major environmental disasters. For example, 2000 died when 20 tons of deadly methyl isocyanate was released from a Union Carbide facility in Bhopal, India. Many CEOs must have questioned whether this could happen to them. The effects of this disaster have been extremely long lasting — on the environment, on the affected population, on the company that was responsible, and on the industry. Six years after Bhopal, an opinion poll in the U.S. indicated that the public acceptability rating for the chemical industry had dropped 25%. It was found that more than 60% "rated the chemical industry as *very harmful to the environment*" (Gunningham 1995).

Nike and Exxon provide more recent examples of how environmental issues can severely damage a company's image. Nike was accused of exploiting laborers overseas by exposing them to substandard working conditions, such as excessive solvent exposure, and in turn marketing expensive shoes to poor inner-city youth. Though the company has made significant strides to ensure that vendors who manufacture their shoes have responsible environmental and safety practices, it seems that this issue never goes away and sticks in the consumers' mind. Similarly, many people have never forgiven Exxon for the Alaskan oil spill, and the anger appears deep-seated. "I think it's going to take another generation for Exxon to recover," said Jeane Vinson of Kona, Hawaii, during a corporate reputation survey. "I can't get the image of dead birds out of my mind" (*Wall Street Journal*, September 23, 1999). The difficulty in changing a negative public perception has been one of the main drivers for the greening of industry.

As industry was trying to protect itself from adverse publicity brought about by incidents and enforcement activity, leading companies started realizing that there were business benefits for being green. Companies started telling stories about how reducing pollution could save money. One of my first encounters with this idea was at a training session dealing with pollution prevention, where I heard about 3M's Pollution Prevention Pays (3P) program. 3M claimed that it had saved significant dollars by preventing pollution, which set some bells off in environmental managers' heads.

> Leading companies started realizing that there were business benefits for being green.

A further step is the advent of sustainability indices and brand image enhancement. For example, by being proactive environmentally, a company can be among a limited number of publicly traded companies that are included in a special stock fund like the Dow Jones Sustainability Index (DJSI), which has been developed for people who want to invest only in socially responsible companies. The DJSI consists of more than 200 companies that represent the "top 10% of the leading sustainability companies." The market capitalization of the Dow Jones Sustainability Group World Index exceeded $5 *trillion*. The index claims to be able to motivate companies to "increase long-term shareholder value by integrating economic, environmental and social factors in their business strategies" (DJSI 2001).

Further, by being a socially responsible company and by voluntarily reducing pollutants, customers feel better about using the services and the public respects those companies over competitors that do not share these ideals. There have been reports that indicate 25% of American consumers claim that they have shifted their buying habits because of negative impressions of a company. Therefore, companies are thinking more about whether they are viewed as an environmentally responsible company (Welford 2000).

> For socially responsible companies that voluntarily reduce pollutants, customers feel better about using the services; and the public respects those companies over competitors that do not share these ideals.

Case histories

In this section, we look at a few leading companies in various manufacturing sectors to see what types of voluntary compliance programs have been initiated. The areas of focus are on what the companies and others have said about their voluntary compliance programs, their compliance performance, and any best practices that are established.

IBM

In 2000 IBM published its 11th annual public environmental report. It was one of the first companies to report publicly on environmental performance

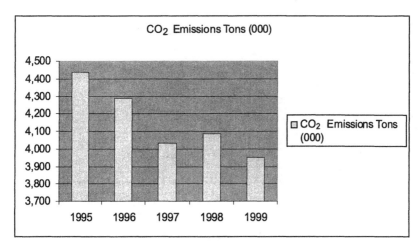

Figure 6.1 IBM CO_2 emission reductions.

and set an example and standard for other companies. IBM claims to have a unique connection with sustainability by providing information via the Internet and e-business capabilities that minimize the use of many natural resources (Lyon 2000).

Voluntary initiatives

IBM has promoted initiatives to reduce the impact of its operations on the environment. A corporate environmental policy states that the company is to be an "environmentally responsible neighbor" in the communities where it operates. They are committed to conserve natural resources primarily through reuse and recycling activities and to develop products that are "protective of the environment." Regarding compliance with regulations, IBM commits to "exceed all applicable government requirements" and "adhere to stringent requirements of our own no matter where in the world the company does business." The company will conduct rigorous audits and self-assessments of compliance with its own environmental policy and regulations" (Lyon 2000).

> A corporate environmental policy states that the company is to be an "environmentally responsible neighbor" in the communities where it operates.

One of the most publicized environmental issues of the day is global warming due to carbon dioxide accumulation in the atmosphere that results primarily from burning fossil fuels. IBM has taken some significant strides to address this issue. Along with Johnson & Johnson, it was the first multi-national company to partner with an environmental advocacy group, the World Wildlife Fund, by joining Climate Savers. Climate Savers is a program

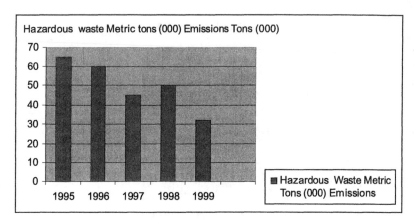

Figure 6.2   IBM hazardous waste reduction.

that encourages energy efficiency and the reduction of greenhouse gas emissions. An important aspect of this program is that participating companies agree to independent verification of their actions (WWF 2000). IBM has already reduced CO_2 emissions by 20% from 1990 to 1997. By signing on to Climate Savers, it agreed to reduce emissions 4% per year through 2004 using 1998 as a baseline (Lyon 2000).

The company has also made impressive voluntary reductions in hazardous waste generation — 45.2% from 1995 to 1999. Even more impressive were the reductions it made from 1998 to 1999 by reducing hazardous waste 32.9% (18,561 metric tons) in 1 year. Of this reduction, 15,483 metric tons resulted from the elimination of a toxic solvent, perchloroethylene, from its facility in East Fishkill, New York.

Other environmental accomplishments of IBM are:

- Reduction of nonhazardous waste by 20 metric tons from 1995 through 1999
- Conservation activities resulting in savings of 4.3 million m³ of water
- Cost savings and cost avoidance from environmental programs amounting to $189,000,000
- A 14% reduction in landfilling of product scrap since 1998 due to its material recovery centers (Lyon 2000).

Compliance

IBM has established several mechanisms to ensure that it is in compliance with regulatory requirements and internal environmental standards. Each site undergoes an annual self-assessment. The corporate internal audit group performs 10 to 12 assessments of IBM facilities against the backdrop of environmental, safety, industrial hygiene, and chemical management, and ISO 14001 requirements. These audits take 3 to 4 weeks each. The results of the audits are communicated to top management, and action plans are put in place to address any unresolved issues (Lyon 2000).

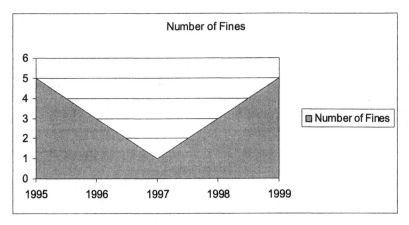

Figure 6.3   IBM fines.

IBM is committed to conform to ISO 14001 management systems. All of its worldwide manufacturing and development sites are covered by a single registration. The company claims that environmental performance has continually improved under this system. IBM voluntarily reports accidental spills and releases in its annual environmental report. In 1999 there were 71 accidental spills and releases reported. While this may seem like a lot, IBM describes its release reporting criteria as exceeding regulatory requirements, and it states that all incidents had little actual environmental impact. Examples of the releases were dilute wastewater solutions, including small quantities of oil and refrigerants. The company also reported it received five fines totaling $9270. The fines were for exceeding wastewater discharge permit limits for chemical oxygen demand from a site in China and improper scrap storage and recordkeeping at a site in Mexico. The latter site has had 17 fines totaling $17,609 from 1995 to 1999 (Lyon 2000).

Best practices

One can learn a lot by studying some of the innovative environmental programs that companies pursue. When companies initiate these programs, the competition in their industrial sector may seek to replicate the program in order to keep pace and not lose a competitive edge. This common practice of "follow the leader" amplifies the effect of the original initiative by influencing companies that sell similar products. IBM has employed several such programs that I classify as best practices.

In 1991 IBM initiated its *Environmentally Conscious Products* program, which they claim "brought about industry-leading practices" in design for the environment. The program focuses on five areas when developing new products:

These tenets provide an excellent basis for any design for the environment program. If IBM's competition is not doing the same thing, they can potentially lose market share because IBM's products are more cost effective to operate and are designed for longer life.

IBM's Design for the Environment Tenets

- Extending product life and upgradability
- Developing products that can be reused or recycled at their end of life
- Insuring that products can be safely disposed of at their end of life
- Using recycled materials in products when possible
- Developing products that are more energy efficient (Lyon, 2000)

IBM clearly benchmarks the environmental profile of their competitors' products; as their annual environmental report states, they are the "industry leader in the use of recycled plastics" (Lyon, 2000). In fact, the company has an aggressive goal to increase the amount of recycled plastics incorporated into their products to 10% of the total plastics purchased by the end of 2001. It increased the use of recycled plastics from 1.13% in 1998 to 6.5% in 1999. Centers have been established that take back old products to manage safe product disposal and recycle as much of the old product as possible. IBM has 14 of these centers throughout the world. This proactive approach will place the company in an excellent position to comply with future take-back legislation that is brewing in Europe and other parts of the world (Lyon 2000).

IBM is a participant in the EPA's Energy Star program. The EPA states that office equipment designed to Energy Star criteria typically uses half the energy of conventional equipment.

The customer also benefits from these environmental policies. The biggest benefit is derived from reduced electric costs from a more energy-efficient product. IBM is a participant in the EPA's Energy Star program. The EPA states that office equipment designed to Energy Star criteria typically uses half the energy of conventional equipment (Lyon 2000).

We have learned that IBM is taking many positive steps voluntarily to protect the environment. It has set a standard in many areas and has led other companies in their industry in a more environmentally friendly direction. It is evident that IBM has good systems in place to stay in compliance with regulations.

Now we should take a look at a company in an industrial group that most often comes to mind when considering poorly performing companies — the chemical industry. We will see through this analysis that there are proactive companies in all industrial sectors.

Du Pont

When we think of Du Pont, we typically think of chemicals; however, the company is much broader than that. It calls itself a "science and technology" company that manufactures high-performance materials, specialty chemicals, pharmaceuticals, and agricultural products. Some of its best-known brands are Teflon®, Lycra® brand spandex fiber, Du Pont Stainmaster® stain-resistant carpet, and Corian® solid surface material (Du Pont 2001). A review of the Du Pont web site attests to the corporation's commitment to environmental concern by a section titled "to do list for the planet." This section expresses the company's commitment to solve some of the world's environmental issues such as turning ocean water to drinking water, making food grow where it cannot, and making high-performance fiber from plants (Du Pont 2001).

A review of the Du Pont web site attests to the corporation's commitment to environmental concern by a section titled *To Do List for the Planet.*

Its Sustainable Growth Progress Report states that its goal is to create shareholder and societal value while reducing its environmental footprint. Their chairman states that the company wants to be a sustainable growth company (Du Pont 1999). Let us explore some of the initiatives of Du Pont and see what type of voluntary environmental programs it has undertaken.

Voluntary initiatives

Du Pont has tried to embrace the concept of sustainability through its corporate commitment. It emphasizes concepts such as providing for future generations while considering its business objectives. It makes its intentions very clear through a public commitment.

The Du Pont Commitment

We affirm to all our stakeholders, including our employees, customers, shareholders and the public, that we will conduct our business with respect and care for the environment. We will implement those strategies that build successful businesses and achieve the greatest benefit for all our stakeholders without compromising the ability of future generations to meet their needs.

We will continuously improve our practices in light of advances in technology and new understandings in safety, health and environmental science.

We will make consistent, measurable progress in implementing this Commitment throughout our worldwide operations. Du Pont supports the chemical industry's Responsible Care initiative as a key program to achieve this Commitment (Du Pont 1999).

One of the concepts Du Pont embraces is that of footprint reduction. Footprint is defined as "injuries and illnesses to our employees and contractors; incidents such as fires, explosions, accidental releases to the environment, and transportation accidents; global waste and emissions; and the use of depletable raw materials and energy" (Du Pont 1999). The corporation has the goal of zero waste and emissions and zero environmental incidents. It has made significant progress in improving its environmental performance and thus reducing its footprint. Using 1990 as a baseline, global air toxins were reduced 64%, and global air carcinogens were reduced 31%. These reductions were achieved while the company increased production 28% (Du Pont 1999).

The corporation has the goal of zero waste and emissions and zero environmental incidents.

The company also has had substantial success with its U.S.-based metrics. Both hazardous waste generation and toxic release inventory releases (TRI; toxic chemical emissions) have been significantly reduced. In addition to gains made for these target compounds, efforts to lower greenhouse gas emissions in the form of energy conservation have paid off. Worldwide energy use remained constant for the last 10 years despite a 28% increase in production.

Besides having the goal of zero emissions and waste, Du Pont established goals for the year 2010 that correlate to its sustainability commitment. To lessen its impact on global warming, it targets sourcing 10% total global energy from renewable sources such as wind and biomass. The company also set a 20% objective for source of its revenues from non-depletable resources like agricultural feed stocks (at 8% currently) (Du Pont 1999).

Du Pont environmental goals

- Zero environmental incidents
- Zero waste and emissions
- 10% global energy from renewable sources by 2010
- 20% revenues from nondepletable resources

There are certain principles to which the company has voluntarily committed that will drive better environmental performance. It has publicly committed to continuous improvement to reduce the environmental impact of its products and processes considering the entire product life cycle. One of its public commitments, which I believe is vital for a company that is trying to be proactive, is to maintain open discussion with stakeholders about its products and their impact on the environment (Du Pont 1999). Perhaps one of the greatest weaknesses of corporations is their failure to explain to the public and interest groups the things they are doing to address environmental issues and concerns.

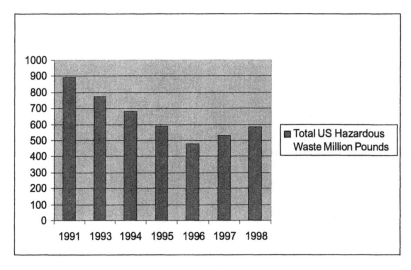

Figure 6.4   Du Pont U.S. hazardous waste chart.

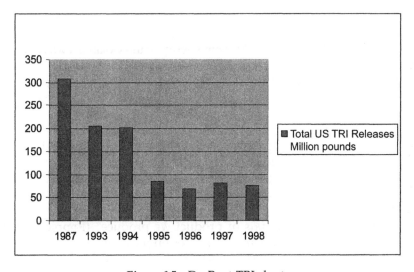

Figure 6.5   Du Pont TRI chart.

Compliance

Like most multinational companies, Du Pont has objectives for compliance with regulations and progress reports. The corporation holds management responsible for ensuring that it is in compliance with environmental laws and regulations and that it provides proper training to employees to meet this commitment (Du Pont 1999). The company is very transparent in depicting its compliance status. A view of its Web site shows current trends regarding fines received. Overall there has been a downward trend in violation fines paid and incidents. Major incidents have decreased to one a year, for the last 2 years, from having incidents of over 100 per year in the early 1990s. Fines have gone down from a

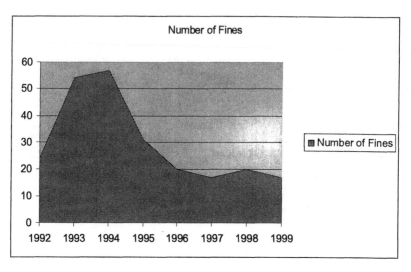

Figure 6.6 Du Pont fines.

high of 57 in 1994 to a low of 17 in 1999. The penalties paid also have been reduced. The company peaked in 1994 with $1.2 million in fines paid and a low of $170,000 in 1999 (Du Pont 2001). The severity of the violations was not clear; neither its Web site nor its sustainability report discussed this issue. However, based on the number of violations and the amount of the fines, at least in recent years, it doesn't appear that there were severe violations.

Best practices

The most forward-thinking practice Du Pont has employed is a metric it introduced called *shareholder value added per pound*. This metric is calculated by the amount of economic value (sales) over the cost of capital over the pounds of final product produced. In the words of Du Pont's CEO, Chad Holliday, this metric will drive Du Pont to "grow our value to society and shareholders without increasing the pounds of product we produce" (Rappleye 2000). This concept is intriguing because it focuses on providing value with fewer raw materials. Therefore, if less material is used to make final product, less material is taken from the earth's natural resources thus reducing environmental impact. This metric is intended to lead shareholders away from businesses that are resource intensive and steer them toward more knowledge-intensive businesses that, in turn, usually result in a greater profit.

> The most forward-thinking practice Du Pont has employed is a metric it introduced called shareholder value added per pound. This metric is calculated by the amount of economic value (sales) over the cost of capital over the pounds of final product produced.

An example of this concept is changing business models from selling pounds of paint to the automotive industry to providing painting as a service. By selling the service, the focus moves away from selling paint to selling painted cars. This results in less paint use and thus less pollution while maintaining a good profit (Coglianese 2001).

I also believe the concept of footprint is a best practice. Typically companies look at their operations in segments: hazardous waste generation, toxic chemical releases, water use, energy use, etc. A footprint is an all-encompassing measure that rolls all of these measures into one number. This doesn't mean doing away with the other measures, which should be subsets to the overall company footprint. It should be the goal of every corporation to reduce its footprint because it depicts the holistic impact of the enterprise.

The review of Du Pont's program has suggested that it is very serious about protecting the environment. The company has voluntarily committed to using nondepleting materials to make its products and using renewable energy sources. It has had good success in reducing some of the indicators of environmental impact such as hazardous waste, toxic chemicals, and carcinogens, and it has introduced some forward-thinking concepts such as shareholder added value and footprint reduction.

Electrolux

Now we will investigate a company that is based in Europe with operations in North America — Electrolux — and see if it has a different approach to addressing environmental issues. Electrolux is a manufacturer of white goods — common household items including washing machines, stoves, and refrigerators. The company is interesting to study because it has made attempts to make environmental issues a core value in its business decision-making process.

Voluntary initiatives

Electrolux looks at environmental programs a little differently than American companies. It came to the realization that, for its business, it can reduce its environmental impact most by greening product rather than primarily focusing on its factories. This point of view is a very prevalent theme throughout its annual public environmental report.

> Electrolux came to the realization that it can reduce its environmental impact most by greening product rather than primarily focusing on its factories.

Electrolux has developed environmental performance indicators. One such indicator measures the effectiveness of sales of its greenest products. The aim is to see more and more of its profit come from green products. An example

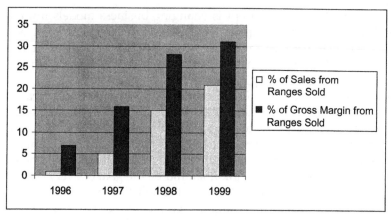

Figure 6.7   Electrolux financial improvement from green ranges.

is what it calls its "green range." The ranges with the best environmental performance had the best financial results for the company. In 1999 the greenest ranges resulted in 21% of the sales and 31% of the profit. It is Electrolux's goal to educate its customers on the fact that they will save money when they buy a more efficient product. The effectiveness of this effort causes the corporation to say, "environmental work generates profitability" (Grunewald 2001).

> In 1999 the greenest ranges resulted in 21% of the sales
> and 31% of the profit.

The company measures the environmental performance indicators of various product types. There is a noted improvement of the sales of its most environmentally friendly products in various types of product lines; therefore,

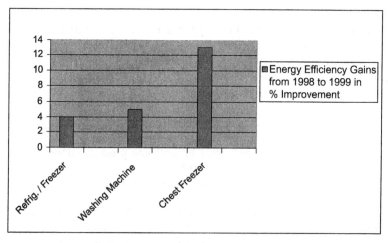

Figure 6.8   Electrolux energy efficiency advances.

the company centers its efforts on improving product energy efficiency. On average, the entire product line yielded improved energy efficiency. Some specific examples of improvements for products sold in Europe are 13% for chest freezers, 5% for washing machines, and 4% for refrigerator/freezer combinations. These green machines not only help the bottom line of the manufacturer, but they also help the consumer. A new refrigerator/freezer combination is anticipated to use a miserly 219 kWh per year. This superior energy efficiency should save the consumer 50 euros per year compared to an average product sold in Europe (Grunewald 2001). With these kinds of results, there will no doubt be pressure put on the competition to keep pace.

Electrolux's American company is Frigidaire Home Products. This company has been selling a washing machine with reduced water use compared with previous models. Considering that 500,000 more efficient machines are on the market, the estimated annual water savings are 2.5 billion gallons (Grunewald 2001). This illustrates the claim that it can improve the environment to a greater extent with its products than its own production facilities.

> Considering that 500,000 more efficient machines are on the market, the estimated annual water savings are 2.5 billion gallons.

This belief has not prevented a focus on improvement of its manufacturing operations. It does a very detailed reporting of the amounts of materials used in manufacturing its products. Each country that has an operation reports on its *inputs* (direct raw materials) and its *outputs* (amount of finished product, waste recycled, incinerated, landfilled, hazardous waste, and emissions to air and water). This gives the public a full picture of the environmental impact of worldwide operations. An examination of the data presented in its year-end environmental report allows the reader to determine the production efficiency when the raw materials are compared with finished products. From 1995 to 1999 the average conversion efficiency of raw materials to finished product was 87%. The data presented also permit a determination whether the company is moving in the right direction at its manufacturing sites. For example, the amount of material recycled has been increasing; however, the amount of materials being landfilled has also been increasing (24.2 tons in 1995, 45 tons in 1999) (Grunewald 2001).

Electrolux reports on a host of other data that the company believes are indicators of its environmental performance — such things as country-by-country usage of coating processes, solvents, and oils, and energy and water usage. Based on data from 152 production facilities, warehouses, and offices, improvements have been reported in energy consumption, carbon dioxide emissions, and water consumption from 1990 to 1999. The way improvement is measured is based on what is defined as "added value" (Grunewald 2001). This is the difference between total manufacturing costs and raw materials

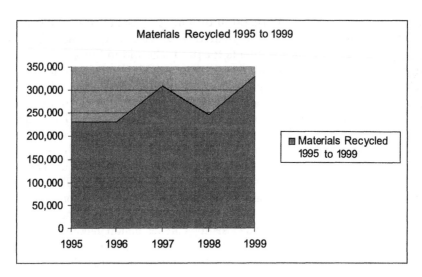

Figure 6.9 Electrolux materials recycled for 98% of worldwide manufacturing sites (tons).

costs. It uses this way of indexing to compare data from year to year because of price fluctuations. It was not entirely clear to me how this indexing is calculated, which makes it difficult for those outside the company to get a good handle on the magnitude of improvement. Putting this lack of clarity aside, it is apparent that there are plenty of data to determine that the company is trying to be responsible and is making a very good effort at quantifying its environmental impacts and improvements.

Compliance
Besides having a commitment to improving its performance through more environmentally friendly products and reducing the impact of manufacturing operations, the corporation has a strong commitment to management systems. The use of such things as environmental performance indicators points to the objectives and targets necessary for a robust management system. Electrolux is committed to ISO 14001; 41% of manufacturing sites were certified by the time the 2000 environmental report was published. Another 30 facilities are in the process of certification; therefore, the company is relying on the voluntary standard to help it meet its environmental commitments (Grunewald, 2001).

Other than its commitment to ISO 14001, the company does not discuss its environmental regulatory compliance status. The insufficiency of reporting compliance data is not typical of corporations that publicly report their environmental performance. The lack of compliance information may be due to its headquarter location in Sweden. Companies based outside of the U.S. seem to focus more on the impact of their products than on regulatory compliance. Therefore, it is not clear how the company performs in this area.

Best practices

A lot can be learned from Electrolux. It has taken some very forward-looking approaches and definitely can be viewed as one of the leaders in green products. The most important initiative that others can learn from is that it has proven green products really can sell and bring significant profit. This was clearly displayed by the green range example discussed above.

Electrolux has proven that green products really can sell and bring significant profit.

Another significant leading practice is a move away from a hard-copy annual environmental report to a "dynamic, continuous forum for information on the internet" (Grunewald 2001). This move raises the bar on corporate transparency by making more information available more often than the annual performance report.

The holistic approach to manufacturing reporting is also a best practice. Traditional corporate performance reports cover the amounts of energy, hazardous and solid wastes, and toxic chemicals used or generated. The Electrolux report covered all of this information and did a worldwide mass balance. This included inputs of raw materials compared to final products and the traditionally reported outputs (wastes and emissions).

Our review of Electrolux's environmental program indicates that the company has taken significant steps to advancing environmental thinking in business. It has initiated several voluntary programs to improve its products and operations.

Now we turn our attention to an industry that is based on making people well — the pharmaceutical business — and look at what Bristol–Myers Squibb is doing.

Bristol–Myers Squibb

Voluntary initiatives

Bristol–Myers Squibb (BMS) is a pharmaceutical company that also markets consumer products and medical devices. Some of the brands that the company markets are Clairol beauty care products; Conevatec wound healing; Zimmer medical devices; and Mead Johnson Nutritionals. The company claims that it has come a long way from focusing solely on compliance with environmental regulations to becoming a "catalyst for a broad range of business, social and environmental benefits." If you consider the company's high rankings by the Dow Jones Sustainability Index, the Hamburg Environmental Institute, and the United Nations Environment program as indicators of being a "catalyst," you would agree with its claims (Hellman 2000).

Many beyond-compliance activities support the strong rankings. One significant initiative was launched in 1992, when BMS was one of the first firms to commit to complete product life-cycle reviews. BMS partnered with the Alliance for Environmental Innovation, a nongovernmental organization focused on improving the environment through innovation, to develop software that integrates environmental considerations into new product development through a product life-cycle review. The assessments evaluate the impacts of BMS products on the environment during their entire life cycle — raw material selection, manufacturing, and product end of life. These types of assessments are known by such terms as design for the environment, eco-efficiency reviews, or industrial ecology assessments (Hellman 2000).

The company has incorporated the assessment requirements into its product development system. BMS has not only improved the environment by making changes, it also *saved more than $7 million through the improvements*. An example of the effectiveness of its design for the environment program is when its Worldwide Beauty Care group reduced alcohol — a volatile organic compound — from 80 to 55% in its Clairol hairsprays. The reformulated products have lower environmental impact while maintaining product efficacy. A further commitment to designing for the environment is its packaging reduction goal. In 1999 the company established a 5-year goal to reduce its packaging by 2.5 million pounds and increase the content of post-consumer materials by 150,000 pounds (Hellman 2000).

Bristol–Myers Squibb environmental goals

- Long-term goal of zero pollution
- Purchase biologically diverse land equivalent to land occupied by its operations by the end of 2001
- Incorporate product life-cycle reviews on all new products by the end of 2001
- Reduce packaging by 2.5 million pounds and increase the use of post-consumer materials in packaging by 150,000 pounds from 1999 to 2004
- Achieve level 2 compliance with the ICC Business Charter for Sustainable Development principles by the end of 2000

In addition to reducing the impact of its products on the environment, BMS has taken a broader view in its quest to be a good corporate citizen. It established a unique goal focused on preserving biologically diverse land. Its goal is to purchase an equivalent amount of biologically diverse acres to the amount of land its operations occupy. An example of this practice is when the company purchased 200 acres of biologically diverse land for preservation in South America in 1998 (Hellman 2000).

Similar to many multinational companies, BMS has focused on reducing the impact of its manufacturing operations. Its long-term goal for operations is zero pollution. A review of the performance information below attests to the effectiveness of its environmental programs. The company tracks its operations' reduction of water, fuel, ozone-depleting substance use, nonhazardous and hazardous waste, and toxic chemical usage on a worldwide basis.

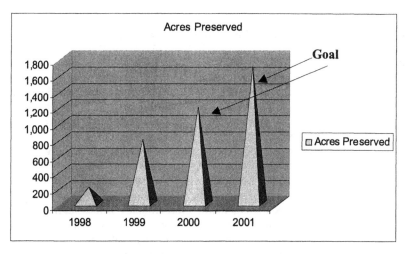

Figure 6.10 Bristol–Myers Squibb biologically diverse land preserved.

Though significant progress has been made, the information presented isn't all good news. For example, hazardous waste increased by 188% in the U.S. from 1994 to 1999 (Squibb 2001). This, however, gives the public more confidence by the company's attempt to be transparent and truthful about its performance.

Bristol–Myers Squibb environmental performance

- Global consumption of water decreased by 46% from 1994 to 1999.
- Global consumption of fuel increased by 3.2% from 1994 to 1999.
- Worldwide electricity usage decreased 36% from 1994 to 1999.
- Greenhouse gas emissions decreased 22% from 1994 to 1999.
- Use of ozone-depleting substances was reduced by 81% worldwide from 1994 to 1999.
- Total nonhazardous waste disposal has decreased by 37% since 1994; recycling of nonhazardous waste increased 66% over the same time period.
- Hazardous waste generation in the U.S. increased by 188% from 1994 to 1999.
- SARA 313 toxic releases to the air decreased by 35% from 1995 to 1999. During the same time period, SARA 313 toxic releases to water decreased 45%, releases to sewer increased 191%, and transfers for off-site energy recovery or disposal increased 112% (Squibb 2001).

Compliance

BMS evaluates its environmental performance through an annual self-assessment process. The assessment is based on the International Chamber of Commerce Business Charter for Sustainable Development — a framework for managing all major aspects of environmental management — which the

ICC Business Charter for Sustainable Development Principles
for Environmental Management

Business activity	Principle
Policy setting	Corporate priority — recognize environmental management among highest corporate priorities
	Prior assessment — assess environmental impacts before starting a new activity
	Products and services — develop products and services that have no undue environmental impact
	Precautionary approach — modify products or services to prevent serious or irreversible environmental degradation
Systems and procedures	Integrated management — integrate ICC principles fully into each business as an essential element of management
	Facilities and operations — operate facilities to minimize adverse environmental impact and waste generation
	Research — conduct research on environmental impacts of raw materials, products, and processes to minimize adverse impacts
	Emergency preparedness — maintain emergency preparedness plans
Implementation and education	Employee education — educate employees to conduct their activities in an environmentally responsible manner
	Customer advice — advise customers and the public in safe use of products provided
	Contractors and suppliers — promote adoption of ICC principles by contractors acting on behalf of the enterprise
	Transfer of technology — transfer environmentally sound technology to industry and the public
	Contributing to the common effort — contribute to development of public policy that will enhance environmental awareness
Monitoring and reporting	Process of improvement — continuously improve policies, programs, and environmental performance
	Openness to concerns — foster openness and dialogue with employees and the public regarding concerns about impacts of operations and products
	Compliance and reporting — measure environmental performance, conduct audits of compliance with the company and the legal requirements, and provide information to employees and the public

Source: GEMI (1994). Environmental self-assessment program, *Global Environmental Management Initiative,* November, pp. 3–5.

company has voluntarily embraced. The ICC principles consist of 16 codes of practice.

BMS evaluates the conformance of each business unit with all of these principles on a 1 to 4 level. Level 1 is the performance that reflects achieving compliance with laws, regulations, and company policies, and Level 4 is continually improving product and processes to enhance efficiencies and competitive advantage. The company-wide goal for all divisions is to achieve a Level 2 status by the end of 2000 (performance relies on management systems rather than individuals to maintain compliance and evaluate products and processes) (Hellman 2000). As you can see from the chart, adherence to the ICC principle is a beyond-compliance commitment.

The company is committed to ISO 14001. As of March 2001, it had 20 sites that received ISO 14001 certification, and many other sites are working toward ISO 14001 certification (Squibb 2001). Some of these sites were the first in their sector to be certified. An example is the Sino-American Shanghai Squibb (SASS) manufacturing facility, a joint venture with the Chinese government, the first pharmaceutical plant in the country to receive ISO 14001 certification. BMS and many other multinational firms have advanced environmental improvements by implementing proactive programs like ISO 14001 in developing nations such as China (Hellman 2000).

The company has a good compliance record that is getting better with time. From 1996 through 2000, BMS received 66 notices of violation and paid 12 fines totaling $201,367 for this 5-year period (Squibb 2001). In 1999 there were two environmental fines issued for minor incidents (a stack test performed in 1997 at a New Jersey site for ammonia on a cogeneration unit resulted in a $6800 fine, and four notices of violation as a result of an inspection of its hazardous waste handling procedures at a site in Puerto Rico resulted in a $4000 fine). There were four small reportable spills in this period, the largest of which was 200 gallons (Squibb 2001).

Best practices

BMS has many initiatives from which other companies can learn. Of the voluntary initiatives discussed, the purchase of biologically diverse land and their design for the environment life-cycle reviews are leading practices. Their partnership with a nongovernmental organization, the Alliance for Environmental Innovation, to improve its life-cycle assessments is also a leading practice.

Further best practices are the efforts made to assist its customers and suppliers with environmental management. BMS has reached out to its customers by providing information and training. In Europe a booklet on the safe handling of cytostatic (anti-cancer) drugs by health-care workers was made available to customers, and a booklet that informs hospitals how they can begin implementing ISO 14001-type environmental standards at their facilities is being made available to customers (Hellman 2000). Further, the company has encouraged "15,000 suppliers and contractors to adopt EHS policies and practices consistent with our own; demonstrate a commitment to environmentally responsible products, services, and management; and align their EHS management systems with the requirements of ISO 14001" (Squibb 2001).

The evaluation of BMS shows how one of the leaders in the pharmaceutical industry is concerned with protecting the environment and voluntarily goes beyond regulatory requirements.

Now we shall evaluate a company that has consistently received high rankings by the public for its good reputation — Johnson & Johnson.

Johnson & Johnson

To determine which companies are most highly regarded or scorned by the general public, Harris conducted an online survey between September 27 and October 17, 2000. Using the results of that survey, Harris calculated a "reputation quotient" based on how respondents rated a company on 20 attributes in six major categories — emotional appeal, products and services, financial performance, social responsibility, vision and leadership, and workplace environment. Johnson & Johnson (J&J) placed first in emotional appeal and products and services. It placed second in workplace environment and social responsibility. Overall the company placed number one in the survey when it was conducted in 2000, and it was also number one in the 1999 reputation survey (Alsop 2001). When the public has this view of a corporation, it is in the firm's best interest to have a very solid environmental program.

> Overall the company placed number one in the survey when it was conducted in 2000, and it was also number one in the 1999 reputation survey.

J&J is a broadly based health-care company with three operating units — consumer, pharmaceutical, and medical devices. The company consists of well-known brands such as J&J consumer products (baby powder and Band-Aids), Neutrogena, Vistakon (Acuvue contact lens), and companies that make medical devices, blood testing equipment, surgical devices, and pharmaceuticals. J&J is guided by its credo, a document that reflects the corporation's principles for the way it conducts business. The document declares that the company has a responsibility to its customers, employees, communities, and stockholders. There is a clear commitment to the environment in its credo:

> We are responsible to the communities in which we live and work and to the world community as well. We must be good citizens.... We must maintain in good order the property we are privileged to use, protecting the environment and natural resources.

Further, since the firm is a health-care company, its environmental mantra links its business objective and environmental objective: Healthy People, Healthy Planet (Larsen 2001).

Voluntary initiatives

I have been a member of J&J's Worldwide Environmental Affairs department for 6 years. I personally feel that my employer is actively trying to be an

Johnson & Johnson Pollution Prevention Goals

Goal	Results
Energy usage — Goal 25% cumulative reduction 1991 to 2000	23.1% cumulative reduction
Packaging usage — Goal 25% cumulative reduction 1992 to 2000	27% cumulative reduction
Water usage — no specific goal	79% reduction
Hazardous waste generation — 10% reduction 1991 to 2000	19% reduction
Solid waste disposal — 50% reduction 1991 to 2000	77% reduction
Office paper usage — 50% reduction 1991 to 1996	64% reduction (1991 to 2000)
Toxic chemical releases — 90% reduction 1987 to 1995	95% (1991 to 2000)
Wastewater disposal — no specific goal	88% (1993 to 2000)
Recycling — no specific goal	64% of solid waste recycled in 2000
Noncompliance events	67% reduction

environmental leader and good corporate citizen. I have included an evaluation of J&J because I can give certain insights into its environmental program beyond what literature can give. J&J has comprehensive voluntary goals that are aimed at reducing the environmental impact of its operations. There are reduction goals for energy usage, waste, packaging, wastewater, and to increase recycling. The goals are aligned with the company's Environmental Affairs Strategic Vision, which commits the firm to environmental leadership.

Johnson & Johnson
Environmental Affairs Strategic Vision

We are committed to environmental leadership, instilling the highest environmental values in all employees, utilizing the best environmental practices in all we do, and focusing on sustainable growth.

The company initiated very detailed environmental goals in the early 1990s, called *pollution prevention goals*, all of which have been achieved. Some of the greatest success has been in the reduction of CO_2, resulting in a 765.2 million pound reduction from 1991 to 2000; a 95% reduction of toxic chemical releases; and a 77% reduction of solid waste disposal (Larsen 2001).

Companies that wish to be considered environmental leaders will have to have a carbon dioxide emissions-reduction strategy.

In addition to having company-mandated goals, J&J is a signatory to voluntary programs like the ICC Business Charter for Sustainable Development and the World Wildlife Fund's (WWF) climate-wise program. According

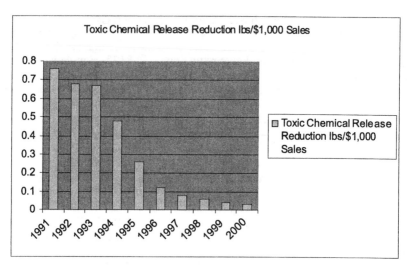

Figure 6.11 Johnson & Johnson toxic chemical release reduction.

to Rebecca L. Eaton of the WWF, companies that wish to be "considered environmental leaders will have to have a carbon dioxide emissions-reduction strategy." J&J committed to reducing its emissions of carbon dioxide, methane, and other gases that are projected to warm the earth's atmosphere by 7% from 2000 to 2010 (Fialka 2000). This will not be an easy task, considering the success it has already had in this area.

The way that J&J reports progress toward its pollution prevention goals warrants discussion. Its 2000 Sustainability report reflects progress toward its pollution prevention goals reported in two manners. Energy and packaging reductions are reported as *cumulative reduction*, and the rest of the goals (water, waste, etc.) are reported as a *reduction in pounds per $1000 sales*. An example of this is noted in the Figure below showing the reduction of toxic chemical releases. The accumulative avoidance for CO_2 shows the amount of CO_2 reduced through the initiation of projects that actually reduced emissions, such as implementation of EPA's Energy Star program. The amounts avoided through actual reductions are accumulated from 1992 to 2000, resulting in 765 million pounds of CO_2 avoided over an 8-year period (whereas the toxic chemical reduction is actual emissions indexed to the sales for a specific year). Therefore, it shows the reduction of emissions per $1000 dollars of sales (Larsen 2001). The net result is that both methods show improvements have been made, which is the objective of reporting the results. However, it is difficult to compare J&J's performance to another company if it does not report the data in the same way.

J&J is moving in a new direction with its voluntary environmental improvement program to what is called *Next-Generation Goals*. This is a new set of goals that has been aligned with the corporation's business goals. For example, a business goal is *manufacturing cost reduction*; the corresponding environmental goals are:

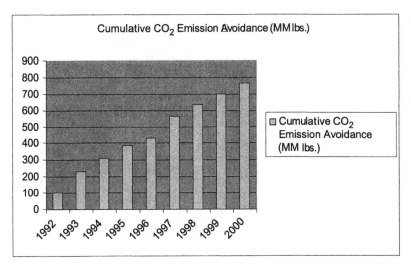

Figure 6.12   Johnson & Johnson CO_2 emission avoidance.

- Energy use reduction (7% greenhouse gas reduction by 2010)
- Water use reduction via implementation of best practices
- Raw material use reduction (packaging and improving the efficiency of converting raw material to finished product)
- Reduction of waste or what they call nonproduct output (anything other than finished product, e.g., hazardous or nonhazardous wastes)

For the business goal of *risk reduction*, the company has a goal of zero violations and accidental releases as well as a commitment to have every manufacturing and research and development site ISO 14001-certified by the end of 2001. For *speed to market* of new products J&J is implementing *design-for-the-environment* thinking in the new product and process development pipeline. To *protect and improve the company's reputation*, site-specific conservation and community outreach programs must be established by the end of 2001 (Larsen 2001).

> This highlights the way corporations are currently viewing environmental goals — goals should be consistent with business objectives and should be a normal part of doing business.

I believe this highlights the way corporations are currently viewing environmental goals — goals should be consistent with business objectives and should be a normal part of doing business. I also feel that more and more companies will take a holistic approach by embracing the footprint reduction concept. This approach embraces the traditional goals of waste reduction and goes further to look at ways to use fewer raw materials in products.

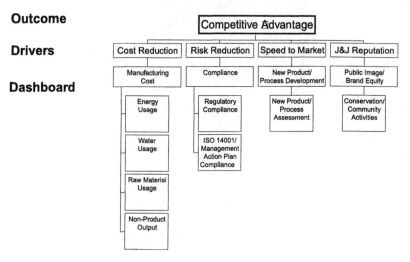

Figure 6.13   J&J business/environmental goal alignment.

Compliance

J&J embraced ISO 14001 to such an extent that it established a goal for all of its manufacturing and research and development sites to be certified by the end of 2001. Considering the company has over 194 operating companies, this will not be an easy task. As of January 2001, the company has had 30 sites certified (Larsen 2001).

J&J employs a self-assessment program to rate the company's status with regulatory and corporate requirements. All manufacturing and research and development sites are audited approximately once every 3 years. Noncompliance events are tracked and reported in its annual environmental report. J&J defines noncompliance events as any accidental releases and regulatory noncompliance incident. There has been a 67% reduction in noncompliance incidents from 1995 to 2000. There is no mention of any fines issued; therefore, one could assume that the events were relatively minor.

Disclosure of this type of information in a public report helps companies to improve performance. No one wants to show a bad trend, especially in regulatory compliance.

Over the past year there has been an increase in events (55 in 2000) compared to 27 in 1999. Eighty-two percent of the events were due to wastewater, and most of these were from the start-up of a new operation.

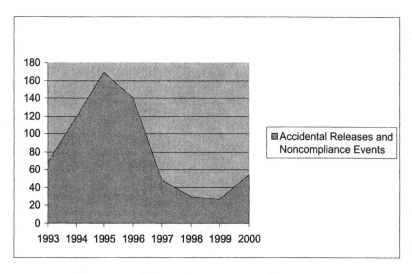

Figure 6.14 J&J regulatory noncompliance events.

Speaking from experience, I can say that disclosure of this type of information in a public report helps companies to improve performance. No one wants to show a bad trend, especially in regulatory compliance, no matter how minor the violation. Merely reporting this data has made a difference at J&J. As can be seen from the compliance chart, overall the trend has been downward. If these data were not measured, I can safely say that there would not be the same level of focus on the results. I personally believe it is important to report all types of violations, even those that seem insignificant. If there is no focus on the small event, it can grow into a huge, highly publicized fine that publicly traded corporations dread.

Best practices

Having a strong value system like the J&J credo is very beneficial to a company in highlighting its responsibility to protect the environment. Having an environmental strategic vision that follows this value system makes a lot of sense. The idea of alignment can be a very powerful one when trying to move a corporation in a more environmentally responsible direction. In my review of corporate environmental reports, I haven't seen many that align environmental goals with the firms' business goals as extensively as J&J does. Focusing efforts on the company's strategic plan can only help a company get buy-in from its management and improve the probability of success. The Next-Generation Goals discussed above are a good attempt at integrating the business and are worthy of listing as a best practice.

> Focusing efforts on the company's strategic plan can only help a company get buy-in from its management and improve the probability of success.

J&J has formed some good partnerships with environmental groups. One such partnership is the greenhouse gas emission reduction pledge taken with the WWF. Another is its work with the Nature Conservancy to protect "critical ecological reserves." J&J has supported projects in Brazil, Mexico, Colombia, and Peru. An example of one such project is the development of conservation and resource management practices in Peru's Amazon basin (Larsen 2001).

Finally, an operating unit in Canada, McNeil Consumer Health Care, has taken an interesting initiative. McNeil calculated the number of trees required to offset its CO_2 emissions. It enrolled 30 local companies to work together and plant 40,000 trees in their community to help offset greenhouse gas emissions (Larsen 2001). The local community has a good opinion of any company that is serious about protecting the environment.

The public expects pharmaceutical and health-care companies to be very responsible and concerned with environmental issues because of the nature of their businesses. We may not expect an industry group that has been associated with air pollution problems to be as proactive. Therefore, we should look at the automotive industry and see what Ford Motor Company is doing.

Ford Motor Company

Voluntary initiatives

One way to look at how Ford views its environmental responsibility is to look for its annual environmental report on its Web site. When you look, you will not find one. This may seem surprising, but the reason is that the annual environmental report — which has been published from 1994 to 1999 — has been replaced by the 98-page *Corporate Citizenship Report*. This title alone gives some indication of the view the company has on its responsibility. To give a further indication of the seriousness of the corporation's commitment, Bill Ford, the Chairman of the Board, makes it very clear in the following quote:

> We see no conflict between business goals and social and environmental needs. I believe the distinction between a good company and a great one is this: A good company delivers excellent products and services; a great one delivers excellent products and services and strives to make the world a better place (Ford 2001).

Ford Environmental Impact 1999

Environmental factor	WW manufacturing facilities
Energy consumption	108.1 trillion BTU
Water consumption	14.5 billion gallons
CO_2 generation	13.5 million tons

Source: Ford (2001). Connecting with society, *Ford Motor Company Corporate Citizen Report.*

This sums up what sustainability is all about — growing business in a responsible manner while helping others.

Of all the companies evaluated, Ford has the largest footprint. With 1.8 million vehicles produced per year and energy consumption of trillions of BTUs, it is operating in a different realm from the average corporation. Ford is very aware of its impact when it admits that its operations and products have "significant" environmental impact (Ford 2001).

Ford has a lot of environmental commitments, starting with protecting the environment, in its core values. "We do the right thing for our people, our environment and our society" (Ford 2001). One of its goals is to take the lead in using recycled material in its vehicles and making the vehicles recyclable. The company has a goal of using more than 132 million pounds of recycled material such as plastic, fabric, and rubber. In 1999 the company used approximately 88 million pounds of recycled materials in its products. Not only is the company incorporating recycled material into its products, but it also uses renewable materials like sisal, hemp, kenaf, and flax. Ford claims that it wants to reduce the bad materials and increase the good materials. Keeping with this theme, it has voluntary phase-out schedules for the toxic metals mercury, cadmium, hexavalent chromium, certain applications of lead, and the chlorinated plastic PVC by 2006 (Ford 2001).

Ford is the first auto maker with plans for a hybrid electric sport utility vehicle. It has a policy called Cleaner, Safer, Sooner that requires the company to put cleaner vehicles on the road ahead of regulatory requirements.

Emissions from vehicles account for 22% of the total CO_2 emissions in the U.S. and 17% in Europe. As expected most of the CO_2 emissions generated by Ford come from vehicle use (70% of lifetime emissions). An indicator of CO_2 emissions is fuel economy. Fuel economy per automobile has improved in the U.S. from approximately 23 to 27.5 miles per gallon from 1980 to 2000 — not ideal for a 20-year period of time. To address this issue, Ford has embarked on an impressive development program for high-efficiency vehicles including innovative technologies like fuel cells, hybrid electric, battery-operated, and alternative-fuel vehicles. Ford is the first auto maker with plans for a hybrid electric sport utility vehicle. It has a policy called *Cleaner, Safer, Sooner* that requires the company to put cleaner vehicles on the road ahead of regulatory requirements. An example of this is selling sport utility vehicles and minivans in the U.S. and Canada markets that meet government low-emission vehicle standards when not required to do so by legislation. These vehicles produce hydrocarbon and nitrogen oxide emissions that are up to 50% lower than vehicles that do not meet the low-emission criteria. Ford had voluntarily improved 2 million vehicles to low-emission status, resulting in eliminating 350,000 full-sized trucks or 4250 tons of smog-forming pollutants (Ford 2001).

Ford goals

- Lead globally by using recycled materials, designing vehicles for disassembly and recycling, and developing an infrastructure for recovering materials from automobiles throughout their life cycles
- Reduce or eliminate substances of concern by 2006 — mercury, cadmium, hexavalent chromium, lead, and polyvinyl chloride
- Implement *Cleaner, Safer, Sooner* policy for putting better environmentally performing vehicles on roads ahead of regulatory timetables
- Provide the best level of emissions possible in countries where there are non-existing or emerging emissions-reduction regulations
- Cut CO_2 emissions in European countries from 1995 baseline in 2008
- Achieve 90% vehicle recyclability by 2012
- Reduce manufacturing energy consumption by 2.25% in 2000
- Maintain ISO 14001 certification for all worldwide manufacturing sites
- Ford suppliers ISO 14001-certified by July 1, 2001 (at least one of each supplier's sites must be certified by the end of 2001)

Another area where Ford has taken a leadership position is its policy in developing nations. It has voluntarily agreed to sell vehicles with the best level of emissions possible in countries where there are non-existing or emerging emissions-reduction regulations.

More emphasis has been placed on evaluating the products' full life-cycle impact. Therefore, vehicle end of life has been getting more scrutiny, especially in Europe. A typical automobile has 75% of its weight recycled. Ford has set a goal of increasing its recyclabiliy to 90% by 2012. This goal exceeds the expected European Union recyclability targets (Ford 2001).

I like the way Ford takes on environmental issues that are a significant challenge. For example, environmental groups have applied significant pressure regarding Ford's SUV Excursion because of poor mile-per-gallon fuel economy. The Sierra Club called it the Ford *Valdez* and the *Suburban Assault Vehicle*. To address this issue, Ford made the case that SUVs are a strength for the company, contributing more profit to the bottom line than any other vehicle. Customer demand drives the marketplace; if Ford did not offer SUVs, it would most likely lose sales and customers would buy its competitors' SUVs, most of which have higher emissions than Ford's. Ford recognizes that the SUV needs to be improved to meet its corporate citizenship goals, so it set short-term goals of increasing the recyclability of the vehicle (the Ford Explorer is 80% recyclable), and all SUVs meet low-emission vehicle requirements — years ahead of any regulatory requirements. For the long term Ford is looking at new methods such as hybrid technology to improve the vehicle (Ford 2001).

The company not only focuses on reducing its product impact but also focuses on reducing its manufacturing plant impact. Because of its big footprint, seemingly incremental change can make a big difference — trillions

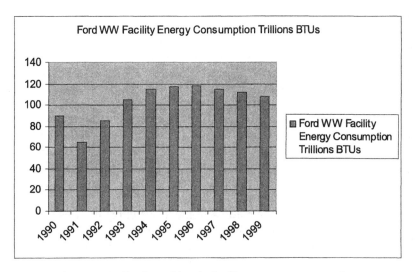

Figure 6.15 Ford worldwide facility energy consumption.

of BTUs or billions of gallons of water. One of its environmental goals was to reduce the amount of energy used. In 2000 there was a target for energy reduction of 2.25% — a 2.4 trillion BTU reduction.

The way it plans to achieve this goal is by pursuing green energy (like solar) and employing energy-efficient technology. The company made a dramatic improvement in water use from 1998 to 1999 by reducing 2.6 billion gallons. There also have been gains in reducing U.S. Toxic Release Inventory emissions. Over the period of 1988 to 1998 they reduced millions of pounds of TRI emissions, achieving a 52% reduction (Ford 2001).

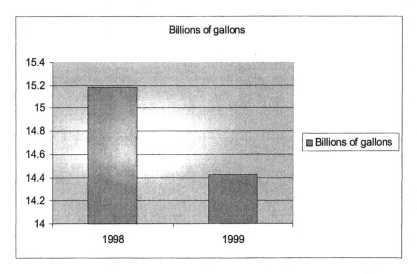

Figure 6.16 Ford worldwide facility purchased water.

Compliance

Ford was one of the first companies that embraced ISO 14001. It has obtained certification for all 114 worldwide manufacturing sites — the first auto maker to accomplish this. The firm's policy is to view regulation as a "beginning point in its environmental stewardship." It has established standards of its own that go beyond legal mandates.

In addition to its ISO commitments, Ford reports its violations from U.S. facilities, regardless of the magnitude. In 1998 it had 58 notices of violation, and in 1999 there were 55. The company received three significant fines in 1999 totaling $1.2 million. The largest was a $1.1 million penalty for air emission violations concerning volatile organic compounds at its truck manufacturing facility in Michigan (Ford 2001).

Though the company has made a tremendous effort with ISO 14001 and should be commended for its truth in reporting, the magnitude of the violations received by Ford highlight an area in obvious need of improvement. It should also be noted that there was no mention of international facility compliance in its Citizenship Report. It is unknown whether there were no violations or whether violations were simply not tracked.

Best practices

One of Ford's strengths is its transparency. It lays all of its concerns on the line in its corporate citizenship report and states how they will be addressed — the SUV issue, regulatory violations, and acknowledgment of a huge environmental impact. Ford has also engaged some prominent environmental leaders and has documented candid critiques of its programs on its Web site. John Elkington, chairman of a leading global sustainability consultancy, and Bill McDonough, cofounder of McDonough Braungart Design Chemistry — a group that encourages sustainable design — have addressed Ford's issues. This approach sends a signal that the company is quite serious about improving its environmental performance.

One of Ford's strengths is its transparency. It lays all of its concerns on the line in its corporate citizenship report and states how they will be addressed.

The company is a leader in implementing ISO 14001 and what is commonly called *greening the supply chain*. Ford is requiring approximately 5000 suppliers worldwide to implement the voluntary management system standard. All manufacturing sites shipping product to Ford must be ISO 14001-certified by July 1, 2003 (Ford 2001). Ford not only makes this a requirement but also offers training to the suppliers. This approach extends the company's environmental protection efforts to a greater extent throughout the world. Ford also extends its improvement programs by using recycled materials in vehicles. Because of the large number of vehicles produced, this fosters the recycled goods market.

The company also can be considered a leader in imposing voluntary pollutant reduction to its products. By implementing new technologies like alternative fuel and hybrid vehicles and by deploying low-emission vehicles into the marketplace when not required to do so, Ford sets an example for other companies to follow.

Analysis of industry programs

Benefits of industry voluntary initiatives

Now that we have evaluated six leading companies, what conclusions can we draw? First, it is evident that there are a significant number of voluntary programs designed to reduce corporate environmental impact. Millions of pounds of pollutants have been removed from the environment. Companies are getting better at reporting environmental results and are opening up discussions with stakeholders. They are calling themselves *corporate citizens* and *environmental leaders* and are trying to lead by example. New ideas are being introduced such as footprint reduction. Corporations are extending their influence to improve the environment by incorporating recycled material into their products; through designing for the environment programs; designing out negative effects; and focusing on product end of life. They are even voluntarily taking back their products for reuse or recycling.

> Companies are getting better at reporting environmental results and are opening up discussions with stakeholders. They are calling themselves corporate citizens and environmental leaders and are trying to lead by example.

Not only are they voluntarily improving their own performance, but they are also extending their improvement efforts to suppliers. Firms are performing self-assessments and auditing their operations to determine their regulatory compliance standing. They are reporting their environmental compliance status and implementing corrective action to improve their compliance efforts. All the firms we evaluated have embraced the voluntary environmental management standard ISO 14001 to help them improve their performance. They all have policies and goals to address environmental issues.

In some cases they are supporting efforts to protect lands that are ecologically significant and even purchasing these fragile environments to ensure they are preserved. Renewable energy sources are pursued, and stringent environmental standards are in place at factories that operate in developing nations that have minimal or no environmental standards. Many have formed partnerships with traditional adversaries — environmental groups.

Similarities

Companies that are environmental leaders all have certain similarities; they have environmental policies and goals that are supported by the top management in their respective organizations. Most firms included photographs of their CEOs in their annual environmental reports with statements of support. Environmental progress is reported at least annually. They have auditing programs to ensure compliance with regulatory and corporate requirements, and they measure performance. All have addressed the impacts of operations and products through efforts to design for the environment. Everyone is committed to continuous improvement of environmental impact and is pursuing the voluntary management system, ISO 14001. In some form or another, each corporation has embraced sustainability principles and has established outreach programs to improve the environment beyond its property boundaries through philanthropic efforts.

What does it take to be an environmental leader?

All companies analyzed had these elements to their environmental programs:

1. Environmental goals supported by the CEO
2. Annual environmental report indicating progress toward goals
3. Commitment to continuous environmental improvement
4. Commitment to ISO 14001
5. Commitment to the concept of sustainability
6. Internal auditing programs for regulatory compliance and company initiatives
7. Support for environmental improvement/preservation efforts through philanthropy

Concerns with industry voluntary initiatives

When something is done voluntarily, it is hard to find fault. Nevertheless, corporations can make more strides in clarifying their annual environmental reports. In several cases it was difficult to understand how performance data were calculated. It was also difficult to compare one company's performance to another because there is no single way to report data. Some companies indexed data to production units or other units, while others reported numbers without indexing.

It would be useful if a standard were developed for reporting certain aspects of performance data. For example, all companies report on the amount of waste they generate. Why not have a standard on how to report these data? Similarly, regulatory violations were reported differently. Some reported all violations regardless of whether a fine was issued; others only reported fines; and in some cases violations were not reported for parts of the world. Standardization will help those outside the company to better understand environmental performance. The standardization in my opinion should only be applied in certain instances as noted above, because companies should always be allowed innovation in their reporting.

Baxter, a health-care company, has taken a step toward adding more validity to its company's performance report. It used the voluntary guidelines for performance reports suggested by the Global Reporting Initiative (GRI). These are guidelines for reporting on economic, environmental, and social performance. Baxter also used an independent company that is an ISO 14001 registrar, ERMCVS, to verify the accuracy of the report. Verification included an "evaluation of Baxter's internal processes for gathering, collating, internally verifying and presenting EHS data to determine conformance with the Good EHS Reporting Principles (GERP) developed by Baxter and others" (Baxter 2000).

This is a step in the right direction. An independent evaluator helps add more credibility to a firm's report. However, the GRI guidelines do not give direction on which performance data must be in a report, nor do they give advice on how the data are to be computed. In the absence of a standard, corporations could improve the descriptions of how data were collected and what the data cover. There should be no doubt to the reader as to exactly how the data were calculated.

Bibliography

Alsop, R. (2001). Survey rates companies' reputations, and many are found wanting. *Wall Street Journal*, February 7, p. B1.
Baxter (2000). *Baxter Sustainability Report*, pp. 1, 50.
Bristol–Myers Squibb (2001). Bristol–Myers Squibb environmental performance. Available: http://www.bms.com/static/ehs/sideba/data/codeof.html. March 18, p. 1.
Bristol–Myers, Squibb (2001). Bristol–Myers Squibb worldwide regulatory compliance. Available: http://www.bms.com/static/ehs/perfor/data/mandat.html# reportablespills. March 18, p. 1.
Coglianese C. and Nash, J. (Eds.), *Regulating from the Inside*, Resources for the Future, Washington, D.C.
DJSI (2001). Dow Jones Sustainability Group Index. Available: http://www.sustainability-index.com/index.html. March 27, p. 1.
Du Pont (2001). Du Pont Corporate Web Page. Available: http://www.dupont.com/corp/gbl-company/overview.html, March 27, p. 1.
Fialka, J. J. (2000). IBM, Johnson & Johnson to pioneer global program to cut gas emissions. *Wall Street Journal*, March 1, p. 1.
Ford (2001). Connecting with society. *Ford Motor Company Corporate Citizen Report*. Available: http://www.ford.com/servlet/ecmcs/ford/index. March 22, pp. 1, 3, 11, 16, 24, 50–52, 70–71, 73–75, 87–90, 93.
GEMI (1994). Environmental self-assessment program. *Global Environmental Management Initiative*, November, pp. 3–5.
Grunewald, P. (2001). Proactive environmental work creates value. Available: http://www.corporate.electrolux.com/files/documents/Environment/Env_report _99_USA.pdf, February 9, pp. 1–2, 4–5, 8–12.

Gunningham, N. (1995). Environment, self-regulation, and the chemical industry: assessing responsible care. *Law Policy*, 17:59.

Hellman, T. H. (2000). *Our Environment Building upon Our Success*. Bristol–Myers Squibb Company, May, pp. 1–3, 5, 10, 12, 15.

Larsen, R. S. and Robert, N. (2001). *Johnson & Johnson Environmental, Health and Safety 2000 Sustainability Report*, Johnson & Johnson, pp. i, 1, 2, 5, 7, 11–12, 14, 26.

Lyon, D. and Wayne, S. (2000). *IBM Environment & Well-Being Progress Report*, IBM, July, pp. 1, 16–20, 23, 32, 94.

Rappleye, W. C., Jr. (2000). From apprehension to comprehension to leadership. *Board*, p. 3.

Wall Street Journal (1999). The best corporate reputations in America. September 23.

Welford, R. (2000). *Corporate Environmental Management*. London, Earthscan Publications Ltd.

WWF (2000). *World Wildlife Fund 2000 Annual Report*, p. 41.

chapter seven

Changing our paradigm

Enforcement of environmental regulation

Much has changed since the start of the environmental movement. Statutes such as the Clean Water Act, Clean Air Act, and Superfund Act were enacted in an era of gross environmental contamination. Air pollution episodes, rivers with dead fish, and uncontrolled waste disposal sites resulting in contaminated drinking water were prevalent. Thirty years later, we are in an era where gross contamination incidents have been significantly reduced.

A complex system of environmental regulations has been developed. Hundreds of pages of federal and state environmental regulations must be analyzed and understood in order to comply.

A greening of industry movement has occurred over the last decade. Companies are voluntarily reporting their environmental performance, setting targets for themselves that go beyond regulatory requirements, and performing internal audits to ensure they are in compliance with regulations and corporate policies.

The EPA is slowly moving away from a command-and-control system and is experimenting with new methods to coax compliance rather than depend on fines, violations, and prosecutions. It has been demonstrated that reliance on enforcement statistics is not a healthy indicator of regulatory compliance.

The full benefit of environmental regulation will not be realized if the regulated community is not following the regulations. Inspections by regulatory agencies improve the adherence to environmental rules. However, the EPA's own records indicate that inspections of regulated facilities are low in number. Regulation without enforcement will not achieve the goal of the regulation.

In the future the EPA and state agencies probably will not have additional resources to enforce regulation and therefore should be doing a better job of focusing their limited inspection resources on problem facilities.

The type of compliance and enforcement program that an environmental agency develops is critical to achieving the goal of regulation. Regulatory agencies have a choice of how they want to view industries. They can look at them as either *amoral calculators* or *corporate citizens*. We discovered that agencies have effectively enticed companies by the use of carrots to coax compliance and with voluntary compliance methods. In many cases, leading companies are doing many good things on their own to improve their environmental profiles like reducing their footprints and greening their products.

Self-regulation has been effectively employed in other areas of society such as financial reporting. Additionally, the current environmental regulatory system relies a good deal on self-monitoring. Initiatives of this sort have value as a regulatory tool; therefore, it is time to consider a new paradigm for achieving environmental protection — one that goes beyond traditional command and control.

> It is time to consider a new paradigm for achieving environmental protection — one that goes beyond traditional command and control.

Analysis of case studies

What did our analysis of the self-regulation and voluntary initiative cases suggest? First we will look at common themes for the benefits of each program, then look at critiques.

Benefits

Improved performance

All five cases indicated that improved performance resulted from each program. The VPP program evidently results in greater protection of workers. OSHA claims that VPP companies are 55% below the typical injury rates of other firms in a specific industry group. Claims of significant cost reductions have been reported due to decreased workman's compensation costs, employee lost-work time, and increased employee morale and productivity (OSHA 1999).

Responsible Care has also resulted in environmental performance improvements. In the U.S. the American Chemistry Council (ACC) claims that the program has resulted in many positive strides:

1. Toxic chemical emissions were down 58% from 1988 to 1997, while member companies' production volumes increased by 18%.
2. There was a 464% increase in the number of community advisory panels formed from 1991 to 1996.
3. Process safety incidents went down from 554 in 1996 to 512 in 1998.
4. Occupational injury and illness rates decreased from 2.93 to 2.13 per 200,000 hours from 1993 to 1999.

5. From 1992 to 1998, energy efficiency improved 13.5% (ACC 2000).

It is hard to debate that the program has not been effective when faced with these data.

In order to be accepted as a XL Project participant, the applicant must demonstrate superior environmental performance. The EPA's Project XL 2000 Comprehensive Report proclaims that some of the main benefits of the program are reduced pollutants including 35,804 tons of criteria air pollutants and volatile organic compounds, increased recycling of 14,006 tons of solid and hazardous waste, and 1846 million gallons of water reused (EPA 2000). There are also cost savings and flexibility benefits for the industries that participate, including reduced paperwork and permit requirements.

StarTrack also has its benefits as detailed in the annual reports submitted to the agency. The company reports evaluated — EG&G, Texas Instruments, and Spalding — all posted significant reductions in pollutants such as wastewater, hazardous waste, nonhazardous waste, and air pollutants. Likewise, all the companies evaluated in the benchmarking chapter posted significant gains in reduction of pollutants through their own voluntary programs.

Beyond-compliance component

All the programs had the characteristic of delivering beyond-compliance results — things that were not specifically required by regulation such as increased reporting to the agency, making more information available to the public, voluntary commitments, and implementing management systems. The aim of the VPP is to encourage companies to go beyond the regulations in cooperation with the agency (OSHA 1996). The program requires statistics and reports that are not required by law to be sent to OSHA. It also forces the participants to abide by more stringent requirements and develop programs that are not specified in the Occupational Safety and Health Act such as management commitment, planning, and accountability.

> All the programs had the characteristic of delivering beyond-compliance results — things that were not specifically required by regulation such as increased reporting to the agency, making more information available to the public, voluntary commitments, and implementing management systems.

Responsible Care has eight fundamental features that include the Guiding Principles and Codes. Most of these requirements are beyond those required by regulation. Examples include reporting on toxic chemical emissions, accidental chemical release incidents, occupational injury and illness reports, and the participation of community members or organizations that ensures the company is doing a good job of protecting the environment. Responsible

Care also requires member companies to report on the level of implementation of the standards and publishes public annual reports.

Participation in Project XL may have various beyond-compliance elements. An illustration of this is the Berry Corporation's project. The company has consolidated 25 environmental permits into one document and committed to superior environmental performance in water use, water consumption and treatment, air emissions, solid waste, and wetland conservation. Some of its commitments included the reduction of 200,000 gallons per day of water and implementation of environmental management systems (EMS) (Freeman 1997). Neither of these requirements are mandated by environmental regulation.

In order to participate in the StarTrack program, a company must commit to several components that are beyond regulatory requirements. First, an EMS consistent with ISO 14001 must be implemented. A company must also perform annual audits of its compliance and EMS systems and divulge its findings to the EPA and state agencies. Additionally, an annual performance report is required for indicating the status of emissions of hazardous waste, air emissions, and solid waste.

The leading companies evaluated had a host of voluntary commitments. These included participation in voluntary codes such as the ICC Business Charter for Sustainable Development; publishing annual environmental performance reports; creating self-imposed emission reduction targets for toxic air emissions, solid and hazardous waste and water use reduction; preserving ecologically sensitive land; and designing environmental considerations into their products.

Public recognition

Public recognition is an important element of each program. The VPP has its Star status, which is conferred with a public ceremony, the use of a logo, a flag to fly in front of the factory, and the prestige that goes along with it. The chairman of American Ref-Fuel has attested to the view that Star status adds the value of public recognition. He claimed that his company became involved with the VPP because it was an "indicator of our company's commitment to safety." The certification builds public trust in the company and helps promote its perception as a good corporate citizen (Siddiqi 1999).

Similarly, the StarTrack program only includes Star companies. The EPA goes out of its way to give recognition for these high performers. Special recognition ceremonies and scheduled press releases ensure that the public knows about the extra effort firms are making in their commitment to the program. Companies covet this type of recognition, and it appears to be the main reason for participation in the program. The information from the Texas Instruments annual environmental report associated the StarTrack program with being an environmental leader (Veale 1988).

Public recognition of a commitment to "doing the right thing" is the primary focus of Responsible Care. An MIT study revealed that most chemical company plant managers thought Responsible Care had radically changed their community relations (Howard 1999). The program also includes the use of a logo, so recognition for commitment to Responsible Care is evident. The very nature of Project XL also gives to the participants a public image as environmental leaders because the companies are working with EPA to shape the future of environmental regulation. Participation in the XL program lends a certain distinction, although not to the same degree as the other programs.

Companies that want to be considered leaders in environmental issues also like the recognition. Signing voluntary agreements to reduce CO_2 pollutants, as IBM and Johnson & Johnson have done, is an example of leadership. It is even better to have an environmental group such as the World Wildlife Fund proclaim that environmental leadership requires a CO_2 reduction policy thus providing public acknowledgment that those companies are leading the pack (Fialka 2000). A high rating by sustainability stock funds and various other environmental ratings, like Bristol–Myers Squibb, is another way to ensure public recognition.

> Companies that want to be considered leaders in environmental issues also like the recognition.

Better relationships with regulators

Perhaps the most frequently mentioned benefit of the three government-run self-regulation programs is the improved relationship with regulators. The best example of improved relations is the VPP program. Reports of agency personnel feeling welcomed at facilities during VPP evaluation visits and OSHA giving advice to companies to improve their programs are prevalent in the VPP. OSHA believes in the partnership aspects so much that it even has conferences entitled "Partner with OSHA" to acclaim the virtues of the VPP. There are many testimonials attesting to the great partnerships OSHA has developed as part of the program (Thomas 1992). The Voluntary Protection Program Participants Association (VPPPA) works just as hard as OSHA to convince industry to be part of the VPP. Also, VPP site personnel are used as special government employees (SGE) to help with the implementation of the VPP program.

> Perhaps the most frequently mentioned benefit of the three government-run self-regulation programs is the improved relationship with regulators.

This type of mutual cooperation is similar to the StarTrack program. The program increases the participant's scrutiny by regulatory agencies because the company must send more reports, tell the agency what's wrong at its site, and indicate how the problem will be fixed. Yet this has resulted in improved relations between traditional adversaries. Regulators have made recommendations for improvements for company programs while on site to witness audits. The program's EMS component has helped the agency learn from one of the participants, EG&G, about ISO 14001 and even allowed the company to perform training sessions and outreach sessions to the benefit of agency personnel (EPA 1998).

Project XL has fostered cooperation between the agency and industry. In order to develop a project, a participant is required to negotiate and work with the regulatory agency. The agency is privy to information about a company's processes that the agency does not normally receive. In the Berry case, the cooperative nature of the program was cited as the biggest success of the project.

Regulatory relief

One of the main incentives for participation in government-sanctioned self-regulatory programs is relief from regulations or enforcement. Project XL is based on "cleaner, cheaper and smarter ideas" (EPA 1998). A site can propose a new way of achieving the goals of environmental regulation that may be more flexible and less expensive. Companies like Merck are able to gain efficiency by obtaining an individual facility-wide (or bubble) emissions cap. This allows Merck flexibility to change its operations without obtaining approvals required by the traditional regulatory mechanisms if they stay below the cap (Wechsler 1998). Regulatory relief in the form of increased flexibility is one of the main reasons companies like Merck participate in Project XL.

> One of the main incentives for participation in government-sanctioned self-regulatory programs is relief from regulations or enforcement.

The VPP allows companies to obtain OSHA consultation in a non-enforcement role. If violations are found by VPP personnel during a verification visit, the company is given the opportunity to correct them without the typical repercussions of an OSHA facility inspection. Participants in the VPP are also taken off the routine inspection program. Likewise, a company in the StarTrack program is given penalty amnesty for non-egregious violations that they discover and report to the agency. There also is the potential to be put at the bottom of the list for inspections. These advantages are not given to companies that do not participate in innovative programs like these.

Criticisms

Now that we have assessed the common benefits of the case studies, we should look similarly at the shortcomings of these programs.

Lack of clarity/overly complex

Perhaps the most frequently cited problems in each case study are unclear and excessively complex program elements. It was evident that Responsible

Care was not as transparent as some would like it to be. As one example, John Ehrenfeld, director of MIT's technology, business, and environment program, claims that the principles of Responsible Care are "far too vague and imprecise to lend credibility to the program and gain the trust of the general public" (Mullin 1998). There is no single, clearly measurable objective toward which the program goals lead. Although measurements in the U.S. are fairly clear and easy to understand, each country can adopt its own measurements. This makes it difficult to compare how the industry as a whole is doing. Additionally, it is difficult to understand all the program elements when reading the Responsible Care Annual Status Report. Undue complexity and vagueness exists in the system.

The most frequently cited problems in each case study are unclear and excessively complex program elements.

Similar critiques have been made of the VPP. The *Revised Voluntary Protection Programs Policies and Procedures Manual* is the 218-page document with which participants must contend. The application process is also difficult, requiring many reports, surveys, and documents. The StarTrack program also has undue complexity and requirements. These include the two separate annual audits (compliance and EMS) and the subsequent reports, summary reports, and corrective action plans. An annual performance report and a third-party certification report are also due every 3 years, with corrective action and follow-up reports.

Project XL has also been accused of being overly vague and complex. Superior performance is not concisely defined. Participants cannot clearly determine the meaning of proper stakeholder involvement. The process is not simple. Extreme amounts of time are necessary to meet the requirements through negotiating with stakeholders and the EPA. In addition, annual corporate environmental reports are sometimes difficult to decipher. It was hard to determine how companies calculated their environmental footprint and to compare one firm's performance to another.

The three self-regulatory programs with government agency involvement all suffered from a lack of incentives for participants.

Lack of incentives

The three self-regulatory programs with government agency involvement all suffered from a lack of incentives for participants. This is not a problem with Responsible Care because its very purpose is to improve chemical companies' images. The VPP is the oldest active government-sponsored self-regulatory program, in place since 1982. Yet despite the efforts of the VPPPA and OSHA, in 1999 there were only 553 companies participating in the OSHA and state versions of the VPP (OSHA 1999). Considering the hundreds of thousands of companies that OSHA must regulate, there is not a lot of participation. We must conclude that the benefits of participation do not outweigh the costs and the many requirements of the program.

Both the StarTrack program and Project XL have similarly low participation rates. As of October 1999 there were only 12 participants in StarTrack (EPA 1999). The participation remains low in spite of EPA's attempts at recruiting more companies. Project XL had a goal of 50 projects; there are only 10 projects that have been approved, with 20 in development as of early 1999 (EPA 1999). Unlike StarTrack, this is a national program, so the lack of participation is even more pronounced. Many have commented on the lack of benefits of the StarTrack program. The EPA even states in one of its evaluations of the program that the costs of participation "generally outweigh current benefits" (EPA 1998). There are similar critiques of Project XL. Industry seems to be taking a wait-and-see attitude about taking part in a voluntary program.

No regulatory standing

Another problem that is common to the three government-sponsored programs is unclear regulatory requirements. This is most evident in Project XL. Fear of participation exists because of the possibility of citizen suits and other legal proceedings against companies with projects. Numerous attorneys have written about the lack of legal power Project XL has to waive permit conditions and other regulatory requirements.

One of the main goals of the VPPPA is to get the VPP codified. The fear is that the program could be cut because it is not clearly called for in OSHA regulations. The group is pushing for a bill to make the program a permanent fixture in national workplace safety legislation because it is funded through OSHA's Compliance Assistance program, which represents less than 1% of the total agency budget. In spite of the fact that it is a small-budget item, the VPP participants are not comfortable with possibly having a budget cut end its existence (Edwardson 1999).

StarTrack also has its regulatory questions. This program's concerns are with third-party certification. The EPA claims that it cannot delegate any "governmental functions to parties outside the agency" (EPA 1998). Because of this view, the EPA's original thought of moving to a system such as the one used by the SEC (with CPAs certifying compliance) is being debated. If those in charge of certification do not have the authority to certify compliance, what can they certify? Therefore, legal issues also must be sorted out in this program.

Stakeholder involvement

Each program has varying degrees of stakeholder involvement. Responsible Care probably has the most stakeholder involvement with the formation of Community Advisory Panels (CAPs) that include members of the community and, in some cases, environmental groups. Even so, there are still claims that Responsible Care is only a public relations program without significant substance. This is probably due more to the sins of the past that developed widespread distrust about the chemical industry than to a lack of outreach; perhaps more can be done in getting stakeholders involved.

The involvement of stakeholders results in the development of unanticipated unity and trust.

There are claims of a shortfall of environmental group involvement in the development of StarTrack. A member of the Audubon Society, involved in a site inspection of a StarTrack participant's facility, tells of the importance of meaningful stakeholder involvement in the development of self-regulatory programs. He feels that if the program is to expand nationally, the public will "need to know that environmental interest groups support it" (Wagner 1996).

Project XL requires stakeholder involvement in its various stages. In many cases the involvement of stakeholders results in the development of unanticipated unity and trust. At other times, as in the case of Intel, results are not as positive. Despite differing views of stakeholder involvement, if meaningful participation can be developed, it will benefit the self-regulatory program by giving it increased credibility and support.

A stand-alone program — one without governmental involvement — will not be effective.

Conclusions from case studies

Now that we have seen that there are some similarities in benefits and concerns with the four different cases of self-regulation, we can analyze the results to see what suggestions can be made about the elements necessary for a successful self-regulatory program. First, the results make it clear that a stand-alone program — one without governmental involvement — will not be effective. We have looked at five different cases. In the benchmarking chapter we evaluated a few leading companies' environmental programs. Though many benefits have come about from companies voluntarily reducing their environmental impacts, improvements are limited to those firms who are interested in being perceived as environmental leaders. Only one program is run by an industry group without government agency involvement: Responsible Care.

As we have seen, Responsible Care is a thorough program with significant elements that have driven some good behaviors: public participation, reduction in chemical releases, safer employees, and reduced toxic chemical emissions. Nevertheless, no matter how strong a self-regulation program is, if it does not have independent checks, it will always have the possibility of

Features of Self-Regulation and Voluntary Programs

Program	Objective	Type of partnership	Key elements	Main benefits	Primary criticisms
Responsible Care	Improve chemical industry perception and performance	Industry initiated and run	Voluntary commitment to RC principles, annual reporting on progress	Improved community relations and improved performance	Vague requirements; no third-party verification; viewed as a public relations campaign
Project XL	Pilot program to test new approaches to environmental protection that demonstrates superior performance	Partnership with EPA, state agencies; and requires stakeholder involvement	Requires site-specific rule making and *Federal Register* notice	Improved environmental protection; more flexibility; lower industry costs	Unclear requirements; legal authority not certain; low incentives to participate
OSHA VPP	Self-regulation of safety programs that go beyond OSHA regulatory requirements	Industry/government partnership	Rigorous application process; commitment to programs more stringent than regulatory requirements	Improved worker safety; cost savings for industry; recognition as a Star company; penalty mitigation	Complex program; increased government oversight; only for stellar performers; inefficient use of OSHA resources
StarTrack	Experimental environmental self-regulatory program	Partnership with EPA and state agencies	Annual regulatory and EMS audits; third-party audit every 3 years; annual performance report	Increased compliance; more data available to public; penalty mitigation; recognition as a StarTrack company	Overly complex; high costs for industry and for agency oversight; not enough incentives to participate
Industry-initiated programs	Be viewed as a good corporate citizen	Typically some form of partnership with an environmental group or voluntary commitment	Annual report; voluntary commitments; commitment to continuous environmental improvement; commitment to sustainability	Perceived as a responsible company	Can be difficult to determine how data were calculated; hard to compare the performance of one firm to another; no independent validation of data

being viewed as a public relations program. Gunningham claims that one of the biggest criticisms of self-regulation as an institution is that it is "self-serving, a concoction of industry groups to give the appearance of regulation when none exists" (Gunningham 1997). The Responsible Care program is on the right track in investigating third-party certification of its programs. If the program moves to independent certification, similar to that of ISO 14001, it will improve its credibility.

We then must look at the three government-run programs to see what can be gleaned from them. Our initial question is what outcome is desired from the self-regulation program? Project XL is intended to test new methods for achieving environmental protection that will yield better results at lower costs. By its very nature, Project XL promotes the development of various types of initiatives that test different elements such as multimedia permits, flexible regulation, and the use of management systems. One of the problems with this program is that it is not clear what will be done differently if any of the experiments work well. In addition, the program is not a pure self-regulation initiative, though it does have several projects that incorporate self-regulation. Although we have learned several things about which self-regulatory elements are useful and which are not, the program in general is not a good model because it tests too many different theories.

This brings us to the StarTrack and VPP programs. Both of these have similar goals, to help accomplish the agency's mission by partnering with outstanding performers to go beyond regulatory standards. There are other similarities as well. Both have standards that go beyond compliance to ensure either increased worker safety or improved environmental protection, although the VPP standards reach farther than those of StarTrack. Each program has an application process that screens potential participants to determine if they qualify for participation. Both have verification programs to ensure that the member companies are adhering to the standards and regulatory requirements. OSHA uses its own resources to perform verification, whereas StarTrack uses third-party certification. Each makes a concerted effort to publicly recognize participants. A model self-regulatory program should have several of the elements from these programs.

Key elements for a self-regulatory program

We can draw certain conclusions from the case studies to determine what the key elements of an effective self-regulatory program should include.

1. Clear goals
   First, the program must have a clear purpose. Is it going to be used to recognize superior performers? Will it improve the focus of limited government agency resources by allowing certain industries to self-regulate? Is it going to test out hypotheses for obtaining better environmental protection? Will it reduce the cost burden of regulated industry? Will only outstanding companies be allowed to participate?

Is the intent to improve performance or just ensure compliance with existing regulation?

2. Government involvement

   To have maximum effect, the program *should be run by a government agency.* It is obvious that public confidence will not be behind a completely industry self-regulated program.

3. Beyond-compliance component

   There should be an aspect of going beyond what is required by the regulations as part of the program. This will enable the agency to recognize the participant as a committed leader and will give the public more trust in the program and yield better environmental results.

4. Clear benefits/incentives

   A good self-regulatory program must be able to make certain promises and deliver on them in order to get and keep participants. Regulatory benefits, whether they are in the form of public recognition, violation amnesty, or reduced inspections, should be spelled out.

5. Minimize complexity

   Every program needs guidelines to ensure that the participants are operating in a manner to achieve the desired goals of the program. However, the elements and subsequent guidelines of a program should be carefully crafted to ensure that they are as simple as possible and should not include any requirements that do not accomplish its goals. This will keep the agency oversight and participant costs at the lowest level possible.

6. Stakeholder involvement

   Stakeholders should be involved in the development of the program to the greatest extent possible so that trust in the program will reach beyond regulatory agencies and the participants.

7. Legal standing

   There should be no doubt that the program is consistent with statutes. There should be no fear on the part of participants that they may receive citizen suits or that funding will be cut because the program does not stand on solid legal footing.

Model self-regulation program

Proposing a new model is the next logical step, now that we have identified the elements that should be a part of an effective self-regulatory program. First, we must define what the goal of the program should be.

Program goal

Based on what we have learned about self-regulation, enforcement, compliance theories, and the case studies, I propose that the goal of a self-regulatory program should be equivalent to the goal of any compliance program: to develop a complying majority. The question is *how is a complying majority developed?*

Key Elements of a Model Self-Regulatory Programs

Element	Requirement	Feature
Clear goals	Program must have a clear purpose	If having a complying majority is the goal, the entire program must be aligned with this goal
Government involvement	A government agency must run the program	Public confidence is greater with government-run programs
Beyond-compliance component	An element of the program should include going beyond regulatory requirements	This will help the participants stand out as leaders and bolster public confidence in the program
Clear benefits/ incentives	Benefits to participate must be obvious and tangible	Public recognition, violation amnesty, and reduced inspections are attractive incentives
Minimize complexity	Program elements must be as simple as possible	There must not be undue burden on the participant or the agency to run the program for maximum benefit
Stakeholder involvement	Stakeholders like environmental groups should be part of the program development	Participation of stakeholders in on-site reviews should be optional because, if mandatory, it will be a barrier
Legal standing	There should be no doubt of the legal efficacy of the program	The fear of legal action is a barrier for participation in self-regulatory programs

We have established that environmental regulation achieves its goal when regulated entities follow the rules. Inspections by environmental agencies ensure that companies are following the rules; however, there are not enough inspectors to visit all the regulated facilities on a regular basis. Therefore, the program goal should be to gain efficiency by focusing agency resources on those companies not complying while building a complying majority with those that are.

> Winning the cooperation of regulated firms is by and large the most effective way of carrying out a regulatory program.

The current system treats all companies with a command-and-control inspection system. As stated by Joseph Rees, "winning the cooperation of regulated firms is by and large the most effective way of carrying out a regulatory program." The regulator should at different times resemble a consultant, politician, or a combination of both. "Regulated firms should be treated as responsible and reasonable citizens motivated by good faith and willing to heed advice" (Rees 1988). The time is right to permit self-regulation of responsible companies in concert with the traditional system. The goal of

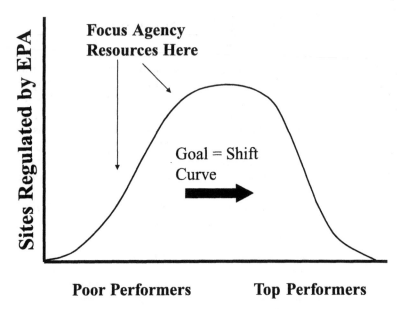

Figure 7.1   Environmental performance chart. *Note:* Goal is to move more companies to the high performing side.

the system is to have more companies become high performers through participation in this voluntary self-regulatory program and allow the EPA and state agency staffs to focus on the poor performers.

Government involvement

The EPA and the state agencies should run the self-regulatory program. The StarTrack model, with its use of a third party to verify compliance, is the most attractive because it expends less agency resources if the agency chooses to infrequently inspect the facilities that participate. The VPP model uses too many agency resources because both verification and program administration are performed by OSHA employees.

Under the model I suggest, agency personnel should only *coordinate* the program, i.e., receive annual reports and process applications for new participants. The agency would reserve the right to inspect the facility to ensure that the site is complying with regulations. However, the inspections would be less frequent than current practice. For instance, if a company in the new model is inspected at the same rate as it is currently, there would be a waste of agency resources since no benefit would be achieved and resources would not be reallocated. Consideration of spot-checking participants at a frequency that is below the average inspection rate is appropriate because the annual reports and third-party verification should be sufficient to give the agency confidence in compliance performance. In this way the agency can focus on the poor performance with the goal of moving them toward better performance as depicted in Figure 7.1.

This approach is consistent with the analysis of self-regulation per-formed by Richard Andrews of the University of North Carolina at Chapel Hill. He states that "third-party certification offers some promise for environ-mental self-regulation where there are clear and consistent standards that can be imposed and verified and where the benefits justify the added costs of engaging commercial certification services" (Andrews 1998).

The main reason for any self-regulating program should be to *get more companies complying with regulatory requirements than the current system does.* Anything achieved above this goal should be considered an added benefit and should not drive the program.

The main reason for any self-regulating program should be to get more companies complying with regulatory requirements than the current system does.

Beyond-compliance component

The program should only be for companies that are interested in recognition for going beyond compliance. In order to recognize a company as a leader, there must be some elements of the program that are above what regulations require, although not so much to make it unattractive. There should also be a way to get companies that might not be superior performers into the program, similar to the Merit track that the VPP has.

The StarTrack program is too limiting because it only allows companies that have a stellar record. The developmental track (or *Merit* as it is called in the VPP) may require additional resources for the agency up-front in order to advise the company. Another way to help develop weaker companies is using the VPP model's mentoring program. Companies that have reached the Star level would be asked to help companies that are in the develop-mental track. This would mean developing an organization like the VPPPA in the environmental arena, thus allowing its member companies to help each other obtain the goals of the program.

The beyond-compliance elements should include annual reports on emissions, a commitment to EMS through ISO 14001 certification, annual compliance audits, and third-party verification every 3 years.

1. Annual report

    The StarTrack program and Responsible Care both require annual reports on performance. Leading companies are providing annual per-formance reports already. We have seen the effect of simply making environmental data available to the public. As an illustration, we should consider the EPA's Toxic Release Inventory. This rule requires reporting releases to air, water, and off-site transfers to waste treatment facilities for several hundred toxic chemicals. All this information is made available to the public. Once news reporters and environmental groups get this information, they report on who is the biggest polluter in the neighborhood. The data availability alone forces industries to control and eliminate these toxic chemicals more than any other command-and-control regulation that I am aware of. The fact is, when this type of data is made available to the public, something must be done to give assurances the company is managing things properly.

The program should have a performance report component that includes the fundamental environmental indicators of a company. The StarTrack model has a very good example of the seven indicators that should be reported:

- Nonhazardous solid waste generation
- Hazardous waste generation
- Water use
- Energy use
- Wastewater discharged
- Air emissions (significant pollutants reported as part of the Toxic Release Inventory or greenhouse gases and ozone-depleting chemicals)
- Significant environmental incidents (reported spills, noncompliance events, any enforcement actions) (EPA 1998)

Most of the leading companies evaluated are providing this information in their annual environmental reports. This type of data, although required for hazardous waste and some air emissions, is not mandated by any environmental regulation and thus is a beyond-compliance requirement. Reporting progress over time for each of these indicators should be an element of the program. Showing charts such as the example indicating tons of solid waste per year (see Chapter 5: StarTrack) should be standardized and made public. Any company participating in this program will want its emissions to look as good as possible; this alone will drive continuous improvement in the seven core elements. As indicated in the StarTrack and Responsible Care case studies, this is exactly what happened.

2. Management systems

Management systems are important to ensure a company's commitment to continuous environmental improvement. Their multimedia approach is a good tool for verifying that an environmental program is managed properly. According to a group of about 25 state environmental agencies called the Multi-State Working Group (MSWG) that is studying the use of management systems, "EMS's should improve compliance with environmental laws, enable organizations to achieve performance beyond compliance with legal requirements, and reduce environmental impacts." The MSWG also goes further and claims that EMS's can be a "supplementary tool" that can be used by regulatory agencies to gain "greater environmental protection" (MSWG 2001).

The EPA also sees the value of EMSs. One of the requirements to join its National Environmental Performance Track program, which recognizes top environmental performers, is that an EMS must be present. Therefore, the agency is stating that in order to be a stellar performers, an EMS must be present (EPA 2001).

One of the StarTrack program's beyond-compliance elements is a commitment to developing management systems. At the initiation of StarTrack, ISO 14001 was not well understood. Now that we have more experience with ISO 14001, we should simplify conformance to an EMS by requiring the participant to be ISO 14001 certified. Commitment to this standard is truly a beyond-compliance requirement because it ensures a systematic method that "addresses the immediate and long-term impact of your company's products, services, and processes on the environment" (Cascio 1996).

ISO 14001

In order to understand the depth of the program and its beyond-compliance elements, we should look closely at its history and requirements. The International Organization for Standardization (ISO) is a federation of national standards bodies from 118 countries. Its purpose is to develop international manufacturing, trade, and communication standards. The American National Standards Institute represents the U.S. Delegates from approximately 50 countries developed ISO 14001 (Cascio 1996). Therefore, the standard was crafted to ensure proper management of environmental issues in a broad way.

The ISO 14001 standard is based on a *plan > do > act > check* cycle. The *planning* stage requires the development of an environmental policy, identification of environmental aspects and impacts and environmental objectives and targets. The *do* requires a company to develop its organization to support the system with documentation, operational controls, and emergency preparedness and response. *Checking* deals with monitoring and measuring, corrective and preventive action, records management, and auditing the management system. Finally, *act* is the management review and monitoring that makes the system continue to work (Cascio 1996). The intent of the standard is to get the entire organization to think about and manage environmental issues. Therefore, it requires a more holistic approach to environmental management — far more than any U.S. environmental regulation. As we can see from the following table, the ISO 14001 elements are broad.

A company can self-declare conformity to the standard or can have a third party perform certification (registration). In this self-regulatory model, the environmental agencies should require certification to participate. This will enable the agency to get out of the EMS business and let ISO 14001 do its job. An independent registrar who is recognized by accreditation organizations such as the Registrar Accreditation Board (RAB) issues the certification. The certification process is detailed and includes an application or

ISO 14001 Requirements

4.0 General	
4.1 Policy	
4.2 Planning	4.2.1 Environmental aspects
	4.2.2 Legal requirements
	4.2.3 Objectives and targets
	4.2.4 Establishing programs
4.3 Implementation and operation	4.3.1 Responsibilities
	4.3.2 Training
	4.3.3 Communications
	4.3.4 Document control
	4.3.6 Operating procedures
4.4 Checking and corrective action	4.4.1 Monitoring and control
	4.4.2 Handling non-conformances
	4.4.3 Record maintenance
	4.4.4 EMS audits
4.5 Management review	

Source: Cascio, Joseph (1996). *The ISO 14000 Handbook*, Fairfax, CEEM Information Services, p. 28.

contract, an initial assessment/document review by the registrar, an on-site certification assessment, and ongoing surveillance assessments.

The purpose of the certification is to attest that the company is adhering to the ISO 14001 standards. The on-site assessments include the review of such items as:

- EMS manual
- Environmental aspects and impacts
- Regulatory requirements
- Audit reports
- Organization charts
- Training programs
- Management review minutes
- Continuous improvement plans
- Interviews of employees

At the end of an on-site evaluation, a report is issued indicating whether the site receives approval, conditional approval, or disapproval. If nonconformities are noted, they are indicated in the report, and the site must submit proof that it has addressed and corrected the deficiency in order to receive full approval. A site that becomes certified will receive a certificate. After certification is obtained, the company enters a surveillance program that includes site-assurance visits, typically once a year but as frequent as every 6 months. Every visit results in a report that indicates the site's status and notes any nonconformities.

Registrars/auditors must have certain skill sets and experience and must maintain certification themselves in order to be in good standing with the

Accreditation Board. They are not allowed to give any advice on how to comply with the standard; they can only give an evaluation of conformance to ISO 14001 (Potts, 1996).

ISO 14001 is a beyond-compliance commitment that has an established infrastructure for assuring conformance. There are EMS standards, requirements for independent third-party registrars/auditors to confirm adherence to the standard, and associated reports indicating conformity. Simply requesting third-party certification gets the agency out of the EMS business and lets the ISO organization take care of it. This eliminates the collection of reports, setting of standards, and qualifications of auditors as the StarTrack program has done. The added beauty is that the agency does not have to pay for anything because the cost of the audits and reporting is borne by the ISO registered company when it contracts with a registrar. This approach minimizes unnecessary complexity.

ISO 14001 is a beyond-compliance commitment that has an established infrastructure for assuring conformance.

Although parts of ISO 14001 require a company to evaluate regulatory compliance, it does not guarantee compliance. For example, the requirement of the environmental policy states that the firm must identify "legal and other requirements to which it subscribes ... that are directly applicable to its environmental aspects." They must "consider legal and other requirements in establishing its objectives and must make a commitment to regulatory compliance." While they recognize that ISO 14001 will not guarantee compliance, regulators such as Brian P. Riedel, special counsel, EPA Office of Planning and Policy Analysis, realize that it should put a company in a better position to "meet or exceed regulatory requirements" (Freeman 1996).

State agencies, such as the Wisconsin Department of Natural Resources, have studied the use of ISO 14001 as a tool. They feel that an EMS is a diligent environmental standard that will "reassure regulator, public and firm that a high level performance is likely to be achieved, maintained and improved upon." They also admit that some states and the EPA feel that the regulatory language of the standard does not go far enough, and they question the qualifications and integrity of some registrars. The MSWG has requested ISO officials to "correct several defects" in the standard to strengthen the regulatory compliance aspects. It suggests changes; i.e., instead of requiring a "commitment to compliance" the EMS should require firms to "obey the law" (Meyer 1999).

Another take on compliance is given by Nash et al. from MIT in their detailed evaluation of the voluntary management system. In their assessment of sites that had ISO 14001 certification, the managers they interviewed felt that ISO 14001's benefit to them was "strengthening regulatory compliance programs." They concluded that it does not guarantee compliance, but it has improved the performance at the sites studied (Nash 2000).

From my experience with ISO 14001, I agree that it definitely strengthens compliance with regulations. The registrars that my company works with review compliance with environmental regulations and will not give out certification if there are any significant regulatory findings during an audit (such as continued permit violations) if the site is not actively working on corrective action. Overall, the ISO 14001 system has driven environmental improvement and has improved regulatory compliance at my company.

Regulatory compliance

This brings us to a fundamental question of this program: *how do we ensure compliance with all the existing environmental regulations?* Assurance of regulatory compliance is where the EPA and state agencies should put a majority of their efforts. We have established that there are already good models for collecting annual performance data and ensuring conformance to EMS standards. Compliance will be assured by having the participant perform annual self-audits by either in-house experts or qualified consultants. Additionally, a qualified third party should verify compliance every 3 years.

The EPA already has put together detailed requirements for auditors and third-party verifiers for the StarTrack program. It seems reasonable to use most of these standards. However, making simplicity a guiding principle for this program, certain changes should be made. The annual audits should not have any requirements that are unreasonable, such as visiting second and third shifts when there are not any unusual operations. The audits should also remove the video and photography requirements. Based on my experience, the EPA and state agencies never visit on the off-shift for routine inspections and rarely take photographs. Why introduce a potentially contentious issue such as videotaping? Also, eliminate auditing for pollution prevention. This is not usually part of routine inspections, and pollution prevention will occur by requiring annual emissions reports. An easy way to obtain consistency of review for the annual audits is to give the companies checklists. All agencies give their inspectors checklists, so why not let the self-regulatory partner use the agency checklists? When I was an RCRA inspector in the early 1980s, it was considered taboo to do this. But what is the point? Do we not want sites to be in compliance? Why not do everything possible to let facilities help themselves into compliance?

Any unbiased professional should be allowed to perform the annual compliance audits as long as he or she meets professional qualifications. Why not allow a professional from another Star company to do the review (as the VPP program allows industry personnel to be part of the OSHA verification team)? The report to the agency should be by exception only. Since the agency is interested in whether there are any noncompliance issues, a company should only report on noncompliance issues and indicate corrective action in the report. Why make this complex with multiple reports? One exception report per year is sufficient.

Every 3 years there should be an annual compliance verification audit. There would be no need to focus on pollution prevention or EMS because the annual emissions report will drive pollution prevention, and ISO 14001 takes care of the EMS. The audit should be performed by an independent professional as indicated in the StarTrack guide. The report should only address regulatory compliance as the annual compliance audits do. The only difference is that company personnel cannot do the audit. A standards board such as the Board of Environmental Auditor Certifications (BEAC) should set the qualifications for the individual. If the EPA does not feel that this or a similar group of standards are tough enough, it should write its own and have any organization that meets the standards allow its auditors do the compliance audits. Analogous to the use of ISO 14001, this relieves the agency of worry about auditor quality and ethics because the certification groups would do it for them. This is similar to the SEC relying on the group that certifies CPAs the oversight of their auditors (Hale 1998). All third-party reports would have to be sent to the EPA and state agencies. The reports should be on an exception basis.

With this model the agency would only have to maintain procedures for applications, annual reports, and annual compliance audits. The only reports submitted would be the annual emissions and audit reports and a third-party compliance verification audit every 3 years. I would recommend that the report be site-specific and not a compilation of many sites. This would pressure each individual site to reduce its environmental footprint. When information is bundled, some individual sites can take a free ride. If the agency wants to see the ISO registrar reports, it can ask for them at any time and the participant would be obliged to make them available. Minimizing reports and the development of procedures by the agency would alleviate complexity and cost for both the participant and the agency.

An important point to stress is cooperation between federal and state agencies. Since state agencies perform the majority of inspections and enforcement, this must be a joint program. The StarTrack program included state agencies in all its developments and benefited from this involvement. Likewise, any effective program must be a partnership of all parties involved.

Incentives and benefits

A self-regulatory program must have benefits that will attract participants. All of the government-run self-regulatory programs have small numbers of participants. We may conclude from the case study research that the companies involved in these government-run programs are in it primarily for recognition. Although they like the idea of other benefits, they participate in the programs regardless of how onerous the requirements are. Consider the costs of a third-party certifier, from $10,000 to $30,000, plus the costs associated with increased reporting and annual self-audits (Kuhn et al. 1998).

The VPP has many costs and difficulties as well. There are numerous additional requirements, reporting, and increased OSHA inspections. Some of the main critiques of Project XL are its complexity and costs. Also consider the number of companies in the programs. Although the VPP is over a decade old, it has only approximately 700 sites. Project XL is struggling to achieve its 50-project goal, and the StarTrack program only had 14 participants. Because of the years of adversarial relationships between industry and government, any environmental program must be very attractive to get good participation.

If there are not sufficient benefits for participants, environmental self-regulatory programs will only attract a handful of companies that are primarily interested in the public relations benefits of being in a proactive program. This would waste everyone's time. If the goal of the self-regulatory program is to gain a complying majority so that the environment will be better protected, it will not be attained if companies see little benefit in participating. The more participants there are, the greater the likelihood of compliance and beyond-compliance activities that will lead to environmental improvement above that of the current regulatory system.

A look at a recent initiative by the New Jersey Department of Environmental Protection called the *Silver and Gold Track Program* shows how important incentives are. This program was developed by an agency that once reported to the new EPA administrator, Christie Whitman, when she was governor of New Jersey. This program has admirable goals — improved environmental results by giving greater operational flexibility to companies that are top environmental performers (NJDEP 2001). There are three tiers — Silver Track, Silver Track II, and Gold Track. Regulated entities are asked to sign binding covenants that require the reduction of pollutants. For example, in addition to committing to ISO 14001 or similar EMS and developing a community outreach plan, Silver Track II firms agree to reduce CO_2 emissions equal to 3.5% over 1990 levels by 2005. Gold Track participants agree to greater reductions for pollutants like CO_2, NOx, VOCs, and hazardous air pollutants.

Only companies that have good compliance records over a 5-year period can participate. In return the department gives participants recognition, a single point of contact for permit applications, expedited permit processing, research project flexibility, and permit application assistance (NJDEP 2001). Gold Track companies receive extra benefits such as priority consolidated reporting via electronic reporting. The DEP has an application in for a Project XL for the Gold Track portion of the program (EPA-XL 2001).

Other than the recognition, one could argue that the incentives given for the Silver and Gold Track participants should be part of the department's normal operation. Also, it is not clear how the agency is using fewer resources for the companies in this program. There are individuals involved in application reviews, someone needs to be a single point of contact for permits, and there are promised benefits like pre-permit application meetings which require the department's resources. This does not seem to result in fewer resources for the agency. A case can also be made that the companies likely to join this program are trying to reduce their emissions anyway, which is the main thrust of the covenant. This is an example of a program that has excellent intentions but will probably only receive minimal participation due to a lack of incentives.

A recent program introduced by the EPA called the *National Environmental Performance Track* is another example of a voluntary compliance initiative that will not reach its potential because of the benefits to participants. The program is "designed to motivate and reward top environmental performance." There are two parts to this program, an Achievement Track and a Stewardship Track. The more rigorous higher level of recognition, Stewardship Track, is still under development and is planned to be released in 2001. In order to join the Achievement Track, participants must:

1. Have an EMS
2. Have accomplished environmental improvement goals in the past
3. Be willing to commit to future improvements
4. Report annually on their performance
5. Have a public outreach program already in place
6. Have a good compliance record

The requirement for future improvements means that goals must be set for at least four aspects in at least two areas from the following list. In the area of energy use, two goals can be put in place, and two in another area. There must be four improvement goals in total.

- Energy use
- Water use
- Materials use
- Air emissions
- Waste generation
- Water discharges
- Accidental releases
- Preservation and restoration
- Product performance

Benefits for participating in the program include recognition as a top performer by the EPA, reduced reporting and monitoring (reporting for various statutes under one report, for instance), and streamlined administrative procedures (e.g., increased flexibility when installing best-available control technology). There were 228 facilities approved as Charter Members of the National Environmental Achievement Track in early 2001. Of the initial group of applications, 25 were rejected by the EPA for participation (EPA 2001).

> I do not think the benefits will cause middle to poor performers to join up. Shouldn't this really be EPA's goal?

Here again, I do not believe the EPA will get a lot of participants. A review of the charter member list shows many familiar names. Similar to the New Jersey Silver and Gold Track, this program is for high performers only. I do not think the benefits will cause middle to poor performers to join up. *Shouldn't this really be EPA's goal?* The company I belong to is a charter member of this program. The reason J&J joined was because it is doing all the things the program requires and would like to be recognized as a top performer. We appreciate the recognition the EPA provides. But will it help the overall compliance and environmental improvement of the nation? Research from the case studies leads to the conclusion that it will be difficult for the EPA to deliver any of the benefits other than recognition.

1. Penalty Mitigation
    What incentives must be offered to gain the largest number of participants? I believe the top benefit must be real penalty mitigation. If we are going to ask regulatory partners to disclose all their non-conformances on an annual basis, we must remove penalties to the greatest extent possible. Understanding this process requires a brief look at the EPA's Audit Policy.
    Regulatory agencies see an advantage in coaxing industry to be proactive and self-regulating. The stated purpose of the Audit Policy is "to enhance protection of human health and the environment by encouraging regulated entities to voluntarily discover, disclose, correct, and prevent violations of federal environmental requirements." The EPA admits that it has limited resources and that "maximum compliance cannot be achieved without active efforts by the regulated community to police themselves" (EPA 1995).
    The Audit Policy (or, as it is called in the December 22, 1995 *Federal Register, Incentives for Self-Policing: Discovery, Disclosure, Correction and Prevention of Violations*) is a tool for the EPA to encourage self-disclosure of violations and better compliance with regulations. The policy reduces or eliminates fines for certain types of violations if they are discovered during an environmental audit and corrected within 60 days (if possible). The punitive part of a penalty that is beyond the economic gain the noncompliance saved the violator (gravity-based penalty) is waived. Criminal prosecutions will not be recommended for voluntarily reported violations. The limits of the policy include the following: if the violation results in economic benefit, a penalty may be issued; however, the EPA states that "insignificant" economic benefit *may* be waived. Violations and repeat violations are not protected by this policy that cause "serious harm or may pose imminent or substantial endangerment to human health or the environment." Additionally, violations of permits, consent orders, and reports that are required to be submitted to regulatory agencies (such as NPDES wastewater discharge permits) are not protected by this policy (EPA 1995).
    Companies have been taking advantage of the Audit Policy. The EPA reports that the first 3 years of the policy caused 318 firms covering 1668 facilities to disclose and correct violations. A survey by the agency indicated that 84% of the companies would recommend its use to

others (EPA 1999). However, if we consider the benefits of the policy, it appears that it would only be in the interest of the company *to report a violation that would likely be caught by an inspector.* Only a few violations, such as releases of toxic chemicals above reportable quantities, are required by regulation to be reported to the agency. Other violations can be corrected when a company is aware of them, and there is no need to report them. Therefore, I believe that the benefits of violation amnesty should go further than the audit policy.

Companies that have minor violations that did not damage the environment or that gained significant economic benefits by noncompliance should be given complete amnesty — not the possibility of having a fine waived. Additionally, only serious repeat violations should be punished. With minor violations, such as not having a container label completely filled out, the participant should not feel it will be subject to a fine. We should consider all the hundreds of regulatory requirements, the complexity of understanding the regulations, and the continuous disclosure of findings before punishing well-meaning firms for infractions that only can be remotely construed as harming the environment.

The benefits of violation amnesty should go further than the audit policy.

---

2. Lowest Inspection Priority

Since the participants perform multimedia audits every year for regulatory compliance, employ EMS, and disclose and correct violations, there is no need to inspect participant sites as often as a site that is not in the self-regulatory program. Consequently, participants should go to the bottom of inspection lists for state agencies and the EPA. At a minimum they should be inspected at a rate less than the average facility. As noted by the EPA, it inspects sites at a rate of less than 1% per year. If the program participants are still inspected at the rate of industries that are not in the self-regulation track, then the program would be considered a *failure* because the environmental agencies will not be focusing their resources on problem areas that are not currently receiving the proper attention.

The EPA inspects sites at a rate of less than 1% per year. If the program participants are still inspected at the rate of industries that are not in the self-regulation track, then the program would be considered a failure because the environmental agencies will not be focusing their resources on problem areas that are not currently receiving the proper attention.

The agencies should *retain the right to inspect the facilities* despite the fact that there will be verification of compliance every 3 years. The inspections, however, should be modeled after the VPP verification visits rather than the typical environmental agency inspection, which has an adversarial spirit to it. There should be acknowledgment of the participant as a *partner rather than an amoral calculator.* Violations discovered during an agency inspection should be treated the same as self-discovered violations; penalties should only be issued for significant violations, just as they are in the VPP.

3. Recognition

Another important incentive for the participants of the case studies is recognition. In fact, this appears to be the main driver for participation in all of the programs studied. The StarTrack, Responsible Care, and VPP programs do this well. Companies believe that there is a public relations benefit for being identified as a Star company. As pointed out by Gunningham, companies covet the use of a logo. Therefore, the self-regulatory program should include recognition of participants and the permission to use a logo on a level similar to the programs mentioned (Gunningham 1999).

A combination of the following three benefits are enough to develop a significant following: (1) violation amnesty for issues that do not damage the environment; (2) placement on the lowest inspection priority; and (3) recognition as a Star company. Greater benefits than these are being pursued by state agencies only for sites that are ISO 14001 certified. A report by the MSWG surveyed the flexibility given to companies that are ISO 14001 certified and indicates that, besides the three benefits mentioned, others are in trial mode. Examples include (1) the state of California is providing streamlined permitting; (2) Iowa has self-permitting for process equipment and modifications, custom reporting, and recordkeeping; (3) Kansas' version of the Audit Policy gives more amnesty for violations self-identified by companies employing an EMS registered by ISO 14001; and (4) the State of Oregon has a four-tiered program that gives greater flexibility to sites that have achieved higher levels of environmental performance (PADEP 1999).

Flexibility is under serious consideration for EMS. I believe the benefits suggested are conservative and will result in better environmental protection because of better compliance. Although other benefits are possible, there are difficulties in implementation. The StarTrack program promised fast-track permitting but was not able to deliver on it. Flexible regulatory requirements like those pursued in Project XL, such as bubble permits, less reporting, and rapid permitting are hard to implement because individual state permit programs are involved, and they have their own internal complexities. One of the goals of the program is to keep regulatory agency oversight to a minimum, so complexity should be eliminated if at all possible. If we choose simple yet well-appreciated incentives, both the agency and industry will realize the benefits. In this way we will attract more participation. Additional benefits can always be added after several years of experience with the program if deemed necessary.

One could argue that corporations should implement a self-regulatory program and reap all the benefits without government's involvement. The truth is that some of the same benefits would result, but not reduced inspections or public recognition. Unless an agency has a program to give these benefits, a company will appear self-serving by stating that it is a leader; if an agency wants to inspect this company as much as a company without a self-regulatory program, it can. Having an agency state that a company is a leader because of its commitments and beyond-compliance activities gives the public more confidence in the firm.

Stakeholder involvement

An important aspect of a national program must include the involvement of environmental groups. Consider the criticism of Project XL by the Natural Resources Defense Council (NRDC). It wanted more balance and consensus-based decision making and also felt that the process was not open (Steinzor 1998). It is worth revisiting the statement by the Audubon society in its comments on the StarTrack program. Audubon agreed with the program, stating that "EPA and the state environmental agencies will never have sufficient staff and budget to do complete reviews of compliance at all industrial facilities on a regular basis." And if the program will allow resources to be focused on industries with "poor compliance records," it will "prove a very beneficial program." But the society added that if the program was to expand, the public would want to "know that environmental interest groups support it" (Wagner 1996).

> Environmental groups are not opposed to self-regulatory programs if they (1) improve the environment and (2) are involved in their development.

In a critique of reinvention programs, Steve Skavroneck of Citizens for a Better Environment gave further insight on what environmental groups' concerns are with reinvention programs. One of the elements he would like to see as part of a new program is a meaningful beyond-compliance effort. "Will the air be more breathable, the water more drinkable, the fish more edible?" He is also right in wanting to know what the agency will *not* be doing as a result of the initiative and what the costs associated with the program are (Skavroneck 1999). These comments suggest that environmental groups are not opposed to self-regulatory programs if they (1) improve the environment and (2) are involved in their development.

The government agencies running the self-regulatory program should try to involve stakeholders as much as possible. Involving them up-front

Model Self-Regulation Program

Program Goal	Program Elements	Incentives	Legal Clarity
Develop complying majority	Commitment to regulatory compliance via annual self-audits and third-party verification of compliance every 3 years	Recognition as a Star company	Third-party does not certify compliance – it *verifies* compliance
Should be as inclusive as possible to attract the greatest number of participants	Annual emission reports for air pollutants, water use, waste generation, and energy use	Violation amnesty for *all* self-disclosed violations except for those that damage the environment or result in significant economic advantage	The agency does not relinquish its regulatory enforcement authority to third parties since inspections will still occur — just at a lower rate for those in the program
The greater the amount of participants, the greater the environmental improvement	ISO 14001 third-party certification	Reduced inspection priority — must be less than the average site inspection rate	Violation amnesty is already a benefit that agencies give for self-disclosure, so no new regulation is necessary

will result in a more effective and widely accepted program. Participation in on-site visits is an entirely different matter, however.

I believe that the proper approach is that of the StarTrack program. Environmental group participation in the actual audits is not a requirement, but rather suggested as a good idea. We have seen throughout our study of Responsible Care, Project XL, and StarTrack that some companies will get stakeholders involved with their site. However, making it a requirement of the program will not be effective because it will prevent maximum participation of industry.

Legal standing

A major concern from the case studies is the legal standing of the self-regulatory program. The *StarTrack Year One Final Report* made it clear that the EPA cannot delegate its enforcement obligations. Notice that the words "verification" and "verifier" are used when discussing the third-party audits. The idea of certification is what muddies the legal waters with StarTrack. If the EPA and the states retain their regulatory compliance role and only rely on third-party auditors to verify that the company is in compliance, there do not seem to be any legal problems. Regulatory agencies can use enforcement discretion to reduce inspections and give amnesty for violations noted and corrected for the self-regulatory companies. This is consistent with what they have traditionally done. All that would be necessary is an agreement between the company and the agency regarding the use of enforcement discretion if the participant agrees to conform to the program requirements.

Conclusion

The Environmental Protection Agency and various state departments of environmental protection were formed to enforce the law. Companies are voluntarily reporting their environmental performance, setting targets for themselves that go beyond regulatory requirements, and performing internal audits to ensure they are in compliance with regulations and corporate policies. Compliance is critical to "realizing the benefits envisioned by environmental policy" (Tietenberg 1992). Having regulation without enforcement will not achieve the goal of the regulation. Agencies are not inspecting regulated facilities on a frequent basis (less than 1% multimedia inspections every 2 years).

The demeanor of a compliance and enforcement program is critical to achieving the goal of regulation. Agencies can view the regulated community as either an amoral calculator, always trying to get around the regulations, or as corporate citizens trying to do the right thing. It is time for a dual system where companies that are trying to do the right thing and are willing to go beyond compliance should be allowed to participate in a self-regulatory program.

We have learned from the case studies that a self-regulatory program must have specific elements in order for it to be effective. There must be *clear goals*. If the program is to get more companies to comply with environmental regulations, then all elements should be focused toward this goal. A successful program must have *government involvement* because the public will not have confidence in a completely industry-run program and will view it as a public relations ploy. There must be a *beyond-compliance component* in order to recognize participants as leaders and to give the public confidence in the program. There must be *clear benefits/incentives* so that maximum participation will occur. We should learn from our past attempts at self-regulation and *minimize complexity* of the program to make it easy for the regulatory agency to oversee the program and for participants to implement the elements. In order to get the widest acceptance of the program, there must be *stakeholder involvement* in its development. The *legal standing* of the program should be clear so that there will be no regulatory impediments to participation.

A model self-regulatory program's goal should be to develop a complying majority. Environmental regulation achieves its goal when regulated entities follow the rules. The more facilities that voluntarily identify their shortcomings and come into full compliance, the more our environment will benefit from the regulations.

The program should be run by the EPA and state agencies in a manner that minimizes their oversight costs. The program must be based on the commitment to comply with all regulatory requirements and have beyond-compliance elements. The beyond-compliance elements should include annual reports on emissions, a commitment to EMS through ISO 14001, annual audits, and third-party verification of compliance every 3 years.

In order to minimize the regulatory agency's resources to administer the program, reliance should be on self-auditing, ISO 14001 management systems, and third-party verification. Participants go beyond compliance as a result of ISO 14001 and annual performance reports. The full intent of environmental regulation is to have fewer pollutants and thus a cleaner environment. This will be realized by the continuous improvement of management systems and the pressure on companies to show good results from annual emission reports.

Considering the low inspection rates of facilities and that the EPA and state agencies do not take a multimedia approach under the current compliance system, *it is hard to see a downside to this program.* Compliance with the current regulations will be greater. Sites will receive more compliance checks than ever before. Companies will have better environmental performance because of the formalized management systems and annual checks by the ISO registrar. More information will be available to the public, and companies will be forced to lower emissions voluntarily so that their performance reports look good. Agencies will have more resources to focus on the bad actors, and companies willing to voluntarily go beyond compliance will be recognized as Star companies.

Bibliography

Andrews, R. N. L. (1998). Environmental regulation and business self-regulation. *Policy Sci.*, 31(3):183.

Cascio, Joseph (1996). Responsible care performance measures. *The ISO 14000 Handbook.* Fairfax, CEEM Information Services. pp. 4–5, 8, 21–28.

Davies, Clarence J. and Mazurek, Jan (1997). Regulating pollution: does the U.S. system work? Available: http://www.cmahq.com/responsiblecare.nsf, September 24, 1999, p. 1.

Edwardson, Antonia (1999). What is codification? *The Leader,* Available: http://www.vppa.org/Leader/Article.cfm?keyArticleID=74, June 23, p. 1.

EPA (1995). *Incentives for Self-Policing: Discovery, Disclosure, Correction and Prevention of Violations.* Washington, D.C., Environmental Protection Agency, pp. 66707, 66709, 66710.

EPA (1998). *Enforcement and Compliance Assurance Accomplishments Report: FY 1997.* Washington D.C, U.S. Environmental Protection Agency, pp. 2-1, 2-2, 6, 10.

EPA (1998). *Project XL Preliminary Status Report.* EPA-100-R-98-008, Washington, D.C., U.S. Environmental Protection Agency, pp. ii, 11.

EPA (1998). Regulatory reinvention (XL) pilot projects. Available: http://www.yosemite.epa.gov/xl/xl_home.nsf/all/fm-4-23-97.html December 17, pp. 1–5.

EPA (1999). *Protecting Your Health & the Environment Through Innovative Approaches to Compliance: Highlights from the Past 5 Years.* Washington, D.C., Environmental Protection Agency, p. 14.

EPA (1999). XL at a glance. Available: http://yosemite.epa.gov/xl/xl_home.nsf/all/xl_glance, August 31, p. 1.

EPA (1999). EPA draft EMS action plan. Available: http://www.epa.gov/EMS/plan99.htm, January 4.

EPA (2001). About national environmental performance track. Available: http://www.epa.gov/performancetrack/program/standard.htm. March 9, p. 4.

EPA (2001). National environmental performance track. Available: http://www.epa.gov/performancetrack/program/standard.htm. March 9, pp. 1–5.

EPA-XL (2001). Project XL final project agreement for the New Jersey Department of Environmental Protection gold track program for environmental performance. Available: http://www.epa.gov/projectxl/njgold/page2.htm. March 29, p. 13.

EPA, Region I (1998). StarTrack Program Guidance Document. *1998 Draft Guidance for Annual Environmental Performance Report*. Boston, Environmental Protection Agency, Region I, pp. 1–3.

EPA, Region I (1998). *StarTrack Year One Final Report*. Boston, Environmental Protection Agency, p. ii.

Freeman, David. J. (1996). Legal and regulatory concerns. *The ISO 14000 Handbook*. J. Cascio, Ed., Fairfax, CEEM Information Services, pp. 389, 391.

Freeman, Jody (1997). *Collaborative Governance in the Administrative State*. Los Angeles, UCLA, pp. 57–61.

Gunningham, Neil and Rees, Joseph (1997). Industry self-regulation: an institutional perspective. *Law Policy*, 19(4363):370.

Gunningham, Neil (1999). *Smart Regulation*. Oxford, Oxford University Press, p. 82.

Hale, Rhea (1998). *The National Expansion of Star Track*. Boston, U.S. Environmental Protection Agency, Region I New England, pp. 10–12, 15.

Howard, Jennifer, Nash, J. and Ehrenfeld, J. (1999). Standard or Smokescreen? Implementation of a Non-Regulatory Environmental Code. Massachusetts Institute of Technology, unpublished paper, pp. 3, 9–10, 14, 19.

Kagan, Robert A. (1983). On regulatory inspectorates and police. *Enforcing Regulation*. Hawkins, K. and Thomas, J. M., Eds., Boston, Kluwer-Nijhoff Publishing, pp. 74–77.

Kirk, Andrea (1997). *Licensing the StarTrack Third Party Certifier: Case Studies of Established Licensure Systems*. Boston, U.S. Environmental Protection Agency, Region 1 New England, p. 3.

Kuhn, Lauren, Langer, Jenn, and Pfeiffer, Amy (1998). *Designing a Provisional System for StarTrack: An Environmental Management Strategy for the U.S. Environmental Protection Agency*. Boston, Massachusetts Institute of Technology, p. 11.

MSWG (2001). Multi-state working group EMSs, environmental performance, and compliance December 13, 1999 final draft. Available: http://ww.iwrc.org/mswgroot/compliance.htm, April 2, pp. 1–2.

Meyer, George E. (1999). *A Green Tier for Greater Environmental Protection*. Madison, WI, Department of Natural Resources, pp. 13–14.

Morelli, John (1999). *Voluntary Environmental Management*. Boca Raton, FL, Lewis Publishers.

Mullin, R. (1998). Critics look for greater commitment. *Chem. Week*, 160(25):39–40.

NJDEP (2001). Silver track guidance document. Available: http://www.state.nj.us/dep/oppc/silver.htm, Available April 2, 2001. New Jersey Department of Environmental Protection, pp. 1–4.

OSHA (1996). *Revised Voluntary Protection Programs (VPP) Policies and Procedures Manual — 8.1a*. Available: http://www.osha-slc.gov/OshDoc/Directive_data/TED_8_1A.html, September 9, TED 8.1a (5/24/1996):14–15.

OSHA (1999). The benefits of participating in VPP. Available: http://www.osha.gov/oshprogs/vpp/benefits.htm, September 23, p. 1.

OSHA (1999). VPP federal program growth. Available: http://www.osha.gov/oshprogs/vpp/fedgrow.html, September 23, p. 1.

PADEP (1999). *Summary of ISO 14401 State Activity.* Harrisburg, Pennsylvania Department of Environmental Protection, pp. 1–17.

Potts, Elizabeth A. (1996). ISO 14001 certification. *The ISO 14000 Handbook.* J. Cascio, Ed., Fairfax, CEEM Information Services, pp. 331–335.

Rees, Joseph V. (1988). *Reforming the Workplace: A Study of Self-Regulation in Occupational Safety Hardcover.* Philadelphia, University of Pennsylvania Press, p. 179.

Repetto, Robert C. (1995). *Jobs, Competitiveness, and Environmental Regulation: What Are the Real Issues?* Washington D.C., World Resources Institute.

Siddiqi, Shahla, and Johnson, D. (1999). Committed to excellence. *Ind. Saf. Hyg. News Mag.* Available: http://www.ishn.com/newsletter/9905/cover.htm, June 23, pp. 1–2.

Skavroneck, Steve (1999). Regulatory innovation: industry and NGO viewpoints. MSWG Web Site. Available: http://www.dep.state.pa.us/dep/deputate/pollprev/ Tech_Assistance/ltconf/skavro.../index.htm, November 3, pp. 1–2.

Steinzor, Rena I. (1998). Reinventing environmental regulation: the dangerous journey from command to self-control. *Harvard Environ. Law Rev.,* 22(1103):142, 143.

Thomas, Lee M. (1992). The business community and the environment: an important partnership. *Bus. Horizons,* 35:4.

Tietenberg, Thomas (1992). *Innovation in Environmental Policy.* Brookfield, Edward Elgar Publishing Limited, p. 22.

Veale, Jr., Francis (1988). *Report on the Environment: A Summary of Environmental Performance.* Texas Instruments Materials and Controls Attleboro-Mansfield Site, p. 31.

Volokh, Alexander and Marzulla, Roger (1996). *Environmental Enforcement: In Search of Both Effectiveness and Fairness.* Los Angeles, Reason Foundation, pp. 1, 3.

Wagner, Louis J. (1996). *StarTrack Program.* Boston, Massachusetts Audubon Society, p. 1.

Wechsler, Benjamin S. (1998). Rethinking reinvention: a case study of Project XL. *Environ. Lawyer,* 5(1):270–272.

Index

Milton Keynes UK
Ingram Content Group UK Ltd.
UKHW040056071024
449327UK00019B/596